# A NOTE ON THE AUTHOR

Brenna Hassett is a biological anthropologist whose career has taken her around the globe, researching the past using the clues left behind in human remains.

Brenna has a PhD from University College. She is a Lecturer in Forensic Osteology and Archaeology at the University of Central Lancashire and is also a Scientific Associate at the Natural History Museum, London. She specialises in using clues from the human skeleton to understand how people lived and died in the past. Her research focuses on the evidence of health and growth locked into teeth to investigate how children grew (or didn't) across the world and across time. Her first book – *Built on Bones: 15,000 Years of Urban Life and Death*, also published by Bloomsbury – was well received by critics at the *LA Times*, the *Guardian*, and *The Times*, which named it one of the top 10 science books of the year. Brenna is a founding member of the TrowelBlazers Project, dedicated to increasing the visibility of women in the digging sciences.

@brennawalks

# GROWING UP HUMAN

## THE EVOLUTION OF CHILDHOOD

Brenna Hassett

BLOOMSBURY SIGMA
LONDON · OXFORD · NEW YORK · NEW DELHI · SYDNEY

BLOOMSBURY SIGMA
Bloomsbury Publishing Plc
50 Bedford Square, London, WC1B 3DP, UK
29 Earlsfort Terrace, Dublin 2, Ireland

BLOOMSBURY, BLOOMSBURY SIGMA and the Bloomsbury Sigma logo
are trademarks of Bloomsbury Publishing Plc

First published in the United Kingdom in 2022. This edition published 2024

A catalogue record for this book is available from the British Library

Library of Congress Cataloguing-in-Publication data has been applied for

ISBN: PB: 978-1-4729-7572-0; eBook: 978-1-4729-7573-7

2 4 6 8 10 9 7 5 3 1

Typeset by Deanta Global Publishing Services, Chennai, India
Printed and bound in Great Britain by CPI Group (UK) Ltd,
Croydon CR0 4YY

Illustrations by Julian Baker

Bloomsbury Sigma, Book Seventy-three

To find out more about our authors and books visit www.bloomsbury.com
and sign up for our newsletters

For all the mothers, and all the fathers, and all the babies.
But especially for mine.

# Contents

## CHAPTER ONE

# Mary, Mary, Quite Contrary: An Introduction

Mary, Mary, quite contrary
How does your garden grow?
With silver bells and cockle shells
And pretty maids all in a row.

We humans are animals – and animals, as we all know, are weird. Everything from high school biology classes and David Attenborough documentaries, straight through to the rather striking performance art oeuvre of Isabella Rossellini,* has shown even the most casual observer of the world that creatures great and small have developed a vast array of techniques for surviving on this planet. Animals burrow, swim and fly; some walk, some run and some sort of squelch along. The great fascination of generations of naturalists has been the plethora of responses to life's ultimate question, loosely interpreted as: how are you going to get through it? Leave it to humans to come up with outside-the-box answers: a childhood that is waving farewell to its evolutionary origins from the back of a cab on the way to who knows where.

You are weird. Your baby is weird and you were weird as a baby. And after that? You just got weirder. And that is for reasons that include planetary domination, but also

---

* If ever in need of a brief break from the monotony of science, watch Rossellini's *Green Porno*. It is safe for the working environment, though not for those easily disquieted by the grim realities of snail sex.

a long tale of penis spikes and hidden ovulation and fertile fat and monogamy and unripe and unready babies and the evolution of dads and the secret purpose of grandmas and some very, very strange parenting practices ancient and modern that make us the miracle that we are: the ape that never grew up. We, alone among the animals, have decided that not only do we want to live forever, we want to be forever young. We take an extraordinary amount of time to grow up. Our childhood is incredibly drawn out – not just compared with other animals but other apes. We are terrible at getting pregnant, then when we do we undercook the baby and end up with a ridiculously helpless infant. We throw that infant off the breast and into a world of toddling terror sooner than any other species, but then we just sort of ... stay kids. For a *really* long time. When other apes are getting down to reproducing, we are still off in Neverland, playing and learning and doing what kids do.

Why should this be? The answer is in our evolutionary history, way out in the badlands of speculation, fossils and other subjects of the science of the past. The weirdness of us is the result of an uncountable number of tiny little decisions made by our parents' parents' parents and so on back for millions of years. To carry a baby or to park it? What kind of milk and how often? Prioritise fat or free-ranging? What to teach the baby and when? Stone tools or Babylonian schools? These types of incremental decisions are still ones parents are making every day and this matters because evolution isn't static. It isn't an arc of progress leading to our sublime selves, perfect and unchanging. Each parenting choice we made – or that was made for us – has resulted in the kind of human we are today. And the choices we make today are the ones that determine what kind of humans we are tomorrow.

The question this book is asking is what that weird, unique human childhood is *for*. What purpose does it serve in the functioning of our societies and our lives? What is the possible adaptive value of teenagers?

Taking the biological stages of growth as a starting point, we can build a framework of how we shoehorn in different adaptive benefits to different parts of our childhood. To do this we need to understand a little bit more about how different species pass on advantage through the generations. There is often a focus in evolutionary thinking on the red-in-tooth-and-claw aspects of adaptation – the cat-or-be-eaten survival-of-the-fittest scenarios are by their very nature pretty arresting, as I'm sure the poor australopithecine Taung child we will meet later in this book could testify, had it not been snatched up and eaten by a very large eagle deep in the hominid past. At a species level, however, death is not necessarily the most critical part of an animal's life. The most critical bit is the ability to grow into a successful adult who can support the continuation of their genetic material; dying is just something that gets in the way. Retraining our sights on the complex processes that underpin our growing up gives us the chance to really interrogate what it is that has made the difference in our species' survival and success.

Throughout this book, we will look at childhood as money spent. Raising a child is an investment: in the future, in those genetic lines, in the propagation of a species or what might one day become a species. More critically, however, raising a child *requires* investment.* There is a clear difference in the allocation of parental resources between the 'releasing a zillion eggs and hoping

---

* So much investment; a really astonishingly unexpected level of investment.

statistics are on your side' versus the slow nurturing of a dependent over the course of years or even decades. But what is this immense, long-term investment we humans are making in our offspring?

Technically, what we offer is an inter-generational transfer of wealth – inheritance – in a variety of forms. This transfer of a differential potential for success down the genetic line is something proposed by prominent archaeological theorist Stephen Shennan* as a deeply fundamental part of how human societies function, even in the halcyon imagined egalitarianism of the past. There are three possible ways to invest in a child; throughout this book we will see how each type of investment has paid off for our species, because this is definitely the one instance where we do *not* do it like the birds and the bees do it.

So, how do we do it? A basic ecological model of investment in offspring that we can use to explain frogs becomes a far more intricate beast when transplanted into animals that have more than one way of passing on advantage to their children. A frog having the stamina to hold on to its mate long enough to get some eggs fertilised requires a very different sort of investment from the one that educated you to the point where you are reading a book on the evolution of childhood – with complicated terminology and an excessive number of footnotes – of your own free will.† What investments human parents can make in their children, and when and how they make

---

* Who also had the misfortune of having to lead the theory seminar the week we got to post-processualism and I discovered that linguistic obscurantism *does* have a place in archaeology.
† Well done, by the way, unless you are somehow being forced to read this under duress, in which case this book is the least of your problems.

them, shapes their life chances for the future. Not only that, but it shapes the societies they live in.

We can think about the potential for investment in the same way a banker does – in terms of wealth or capital. You do not have to look very hard to come across a form of wealth commonly ascribed to almost all living animals – the kind that they wear. Traditionally, we describe investment in a baby animal in terms of the energetic costs incurred by the parent,[*] as calories diverted to growing the baby. Thinking about 'fast' versus 'slow' reproductive strategies, we can see with a very rough sort of calculation of effort-per-child the differentiation between the shotgun approach to birth and the 'get-my-shotgun' approach to offspring care that appears in humans. The investment a parent makes in this sense is physical, and it is manifested in the actual body of the child that has been grown at such effort; we call it 'embodied' capital. Calories diverted towards building tissues, organs, nerves, etc. are all investments in embodied capital, made by the parent to ensure the viability of their offspring. This is proposed to go beyond the simple business of building an operational child, however.

Embodied capital also includes the *function* of that physical body. Coordination, flexibility, strength – these are all critical to not only the survival of the child, but its success, and they are all built on bone and flesh. To have the physical capacity to survive and reproduce is to possess embodied wealth – and it is not equally distributed. Better-provisioned, better-built animals will be more able to compete for resources and mates, and they will then pass

---

[*] By which we usually mean the cost to the egg layer, but can also include the cost to the parent that has to attract the egg layer and fertilise the eggs as well.

those advantages on to their offspring – who in turn will be better off. A child who has been well fed, its limbs all preserved and motor functions intact, is an example of embodied wealth. We will see in the chapters that follow how humans have developed a unique way of diverting food and resources into finding a mate, producing a baby and fattening it up.

There are other ways of investing in offspring, however, and other forms of wealth that can tip the scales towards species success – and these are the levers that humans have grabbed as we stretch the boundaries of what is possible in terms of investing in the next generation. Part of learning to live in the world is learning to live with the people in it, and for social animals this is sufficiently advantageous to survival that it has become a major focus for investment. This is investment in social capital, and while it may seem the amorphous domain of human celebrities and social media influencers, social capital is a very real asset – and if you don't have it, you don't do well in a society. Crows, for instance, are very big on society. They live in groups and therefore have to somehow figure out how to exist among a bunch of animals known collectively as a 'murder'. To do this, they have evolved an amazing ability not only to recognise each other, but to pick out the social position of other crows in a crow-fight-crow world. Researchers figured this out by staging a sort of knockout tournament between more- and less-dominant crows, and getting another crow to observe them. A crow could easily pick a winner when it knew that one crow was high level and the next crow along was low; it could still pick the winner if it only knew that one of the crows had lost or won a fight to a crow it knew the rank of. Even without theorising the existence of crow bookies, it's clear that this understanding of relationships is sufficiently important to crows to invest in. A baby crow needs to know how to keep its place in the group in order to take advantage of all that group living provides – protection, support with offspring and where the next meal may be coming from. A crow that excels at getting along in its murderous society will have a host of social relationships to call on when, say, a hulking big raven steps in on its territory, and the ability to see off such competition depends on social capital.

Social capital is built on the back of what Monique Borgerhoff Mulder and colleagues called 'relational wealth': the social ties an individual – they mean among humans, but it applies to crows too – can call on for aid in various tasks. These ties might be blood ties, but they might also be formed through other types of socialisation like play or trade exchanges that over time build up social capital. Building up this capital – learning how to be a social crow – requires a type of investment that goes beyond mere care and feeding. It requires social learning. Social learning is the complex, time-consuming method of conveying information down through generations and across societies that provides the training that instinct cannot. From teaching simple lessons like which foods to eat to complex skills like tool-making, many animals up and down the complexity scale of life invest in a type of information transfer that confers adaptive advantage to their offspring – and we are no different. As a matter of fact, we are so dedicated to social learning that we have devoted an unprecedented amount of time to the stage of life where we get most of our learning done: childhood. We will see later on in this book how primates play – and how our species has taken the serious business of child's play and used it to reshape our social worlds.

The final form of wealth we might think of in terms of investment in offspring is, of course, actual capital. This would be material wealth, which can be converted into evolutionary advantage easily enough through enhanced survival or reproductive success. Humans are, thus far, the only animal that has taken this last and final step in the means of generating an extra edge against their fellow humans. No matter how excellently squirrels hoard, they will never manage to turn their nuts into the ability to leverage more nuts down the line. They can convert those nuts to physical benefit, and grow bigger and stronger and better squirrels, but they have not yet worked out how to use their hoards to lord it over their neighbours.* Not even our cleverest primate relations have worked out ways to turn caches into bankable assets,† though, as with all things tripping into the economic sphere, you could add a bit of sophistry and wonder if primate behaviours like meat-sharing, territorialism − and the inherited hierarchical social structures that underpin them − might not find corollaries among bankable human strategies for success.‡

By and large, however, when we are talking about material wealth, we are talking about a type of investment particular to humans and, more specifically, relatively recent humans. The kind of wealth that buys you fancy parenting books, organic cotton baby slings and top-notch therapy for your existential angst. We have created a

---

* Squirrels being notoriously bad at utilising offshore tax havens.
† Though one suspects the macaques of at least considering it.
‡ Indeed, there is an entire genre of economics and organisational behavioural texts that purport to do just that. One enterprising former primatologist has even teamed up with zoos to offer a (copyrighted) course in 'Apemanagement' that lets wealthy, powerful businesspeople watch primates to improve their managerial style. One wonders what, exactly, these captains of industry might make of the bonobo's reliance on promiscuous, non-reproductive sex in relieving social tension. I imagine HR has to sign a waiver.

childhood for our species that we are so unsure about we can't think what to do but throw money at it — unless, actually that is our entire evolutionary strategy. It may sound strange, but if we follow our steps back carefully, we can actually see how investment — and the choices we have made about when and where to do it — have shaped the species we are now. This is the final point for consideration, as we cruise through the evolutionary history of our species that got us the strangest, longest, most-invested-in childhoods on the planet: how much further can we go?

Throughout this book, we will be looking at the paths taken and not taken, from our shrew-like primate ancestors back in the Triassic to our recent relatives, in investing in our offspring. By carefully mapping out the evolutionary possibilities, we can see where we've moved a lever here or dialled down there, slowly creating a childhood that is uniquely, gloriously ours. From fat, easy-cook babies to forever children, every single step on the way to the childhoods we have today has been taken through the process of evolution and adaptation. If we want to know why, we can compare the way we invest — and when we invest — with not only our primate clade, but all the wild and woolly options available to the animal kingdom. It's in this comparison that we see how cleverly we have chosen to invest in our children — and how it has led to the strange and wonderful childhood that has allowed our species to succeed beyond a little Triassic tree shrew's wildest dreams.

## CHAPTER TWO

# Pop! Goes the Weasel: Life History and Why it Matters

All around the mulberry bush,
The monkey chased the weasel.
The monkey thought 'twas all in fun.
Pop! Goes the weasel!

Long ago, in a world far away, a hero arose. More specifically, on Monday, 21 September 2015, in New York, a rat inspired the world.* This particular rat became an internet sensation after being filmed dragging a slice of pizza twice its size down the steps of the New York City subway with the kind of patient determination you just don't expect to see in rats these days. The rat had doggedly – rattedly? – taken a corner of the slice in its jaws and was dragging it down, tugging to pull it across each step then leaping, pizza still firmly clenched between its teeth, into the abyss that was the next step, occasionally tumbling but never giving up. An individual slice of pizza is not, it transpires, an easily transportable shape, but there it was, trying its best. It was enough to earn the rat a brief flare of internet fame and a spot opening a book that is supposed to be about humans.

This book is about humans. As Carly Simon noted in the 1970s, we're pretty vain, and this applies to our species in general as well as to the villain in the chorus who probably thinks this song is about him. But in this case, we're right

---

* Watch it here – youtube.com/watch/UPXUG8q4jKU. That the entire thing was possibly staged by a performance artist called Zardulu only makes it more relevant to the story of humanity.

to be. I am talking about animals because the way we understand humans is through the much broader lens of ecology, and the patterns – or lack thereof – that we find in animals that are more or less like us. And what are we like? Well, apes, really. We are a lot like apes. Which are a lot like primates, which are a lot like mammals, which are a lot like vertebrates, and so on and so forth until you end up in the chorus of a Joni Mitchell song.* But what characteristics of our larger family tree have any impact on what it means to grow up human?

We don't talk about weasel childhoods or toddler turtles, but perhaps we should, because there is a pattern to the growth and development of species. There are any number of ways to schedule each aspect of the business of making more creatures and the differences in these developmental trajectories mean the difference between whole ways of being; whether you are born ready, or just born squidgy and helpless. As a concept, life history is very simple: you break down each part of the life cycle and work out how long it takes for a species to get from stage to stage – from birth to sexual maturity and to the end of their reproductive life. In practice, it's a complicated formula of species trade-offs in terms of how many offspring per birth, how big or developed these offspring are at birth, how many of them are likely to survive and how much to invest in those offspring.

So, to begin, we will look at the way we understand how animals grow up, the framework that scientists use to mark out all the key stages of life that determine how we grow: how long we gestate, how long we depend on our parents, how long we take to get big, how long we take to reproduce and how long we take to die. These are the

---

* Yes. We *are* stardust.

components of life history and the pieces that make up our life histories are not utterly random. They have a connection with factors such as the size of an animal, the likelihood of survival to adulthood, and just how much effort it takes to grow an animal to completion from how they start out in life. We will look at a few different examples from the animal kingdom of how lives might be arranged, and what evolutionary and ecological pressures might push a creature one way or another. But, don't worry. This song is definitely about you.

Broadly, life histories come in two flavours. In a fast life history, animals live fast and die young, and leave a lot of babies behind. Slow life histories are the opposite. Different patterns of longevity reflect different evolutionary strategies, separating animals that opt for a shotgun approach to reproduction from those that invest heavily in a smaller number of offspring. This is easy to see in practice – rats have a *lot* of babies that take less than a month to grow, whereas giraffes take 15 months to cook up a single calf and that calf takes another seven years to make more baby giraffes. In rats, the total investment per offspring is relatively low, perhaps a few slices of pizza; in giraffes, well – you can imagine how much it costs to feed a giraffe.* The options here range from the live-fast-and-die-young James Dean school of thought to the live slow, die old philosophy of nearly everyone else.

What tips an animal one way or the other? The biologist Eric Charnov predicted that larger animals live longer, but also have a longer period of immaturity – they require more growing to get where they're going. The bigger the animal, the longer it lives; we could more or less take the

---

* Apparently, it costs about $3,000 to feed a giraffe for a year.

weight of a given animal and be able to predict lifespan.[*]
This makes sense to us – a big animal like a giraffe takes a
lot longer to grow than a small one like a rat. If that's true,
we should be able to predict, with a careful assessment of
the size of the adult animal and the energy it needs to get
there, both total lifespan *and* how long that animal spends
in what we call, in humans, childhood. The bad news is
that while body mass is a major factor in determining
whether an animal has a fast or slow life history,[†] this is not
a straightforward equation. Animals have different strategies
for how they use that time. There's the size of the adult
animal you're trying to grow, sure, but there's also the size
of the baby, and the amount of growth to be done before
and after birth; not to mention working out if your baby
animal is even going to make it to contributing to the next
generation. This means deciding on an investment strategy
for how you make the next generation. Evolutionarily,
having a lot of babies is a good idea when there's going to
be a lot of them that don't make it and carefully raising
them is probably not a great use of time or energy; you can
afford to go slow with fewer babies when it's safer outside.
Biologists originally formulated[‡] this part of life-history
research as the difference between animals in an unstable

---

[*] Mammals – we're talking mammals. No room here for biological
or spiritual debates about the eternal life of tardigrades.

[†] If you were to decide, for reasons of your own, to unroll animals
and measure them as a two-dimensional surface, the total area of an
adult human weighing about 70kg is 18.361m; a white rat of about
30g might have a surface area of 40cm.

[‡] The formulation of $r$ (mortality) and $K$ (fertility) selection is based
on Pierre-François Verhulst's 1838 equation of population dynamics
that, while comfortably mathematical-looking, is no longer
considered up to the job of describing the full gamut of population
possibilities.

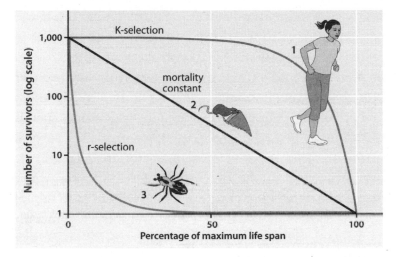

Figure 2.1. *An example of r versus K life strategies. A barn spider has a short life span (approx. 1 year) and very low survivorship at birth, while Pizza Rat and humans make different trade-offs in terms of survivorship versus life span.*

environment that was likely to chew up a lot of babies – where mortality ('*r*') is the main pressure – and those in a stable environment, where fertility ('*K*') is more important.

The pressure on the animal's reproductive pattern from the risk of the babies dying can predict whether animals invest in one birth of a highly dependent infant at a time, or in slightly fewer dependent infants in litters, clutches or even something called a clutter. The 'clutter' is the more dramatic approach of spawning hundreds of babies at a time like Charlotte the barn spider, doomed heroine of *Charlotte's Web*, who dies shortly after laying and preparing her eggs. Luckily for the next generation of spiderlings, they don't have very long to get weighed down by the tragedy of it all – with a 1 per cent survival rate, very few of them will make it, and those that do won't last more than a few years. For Charlotte and her ilk, the shotgun approach is necessary because you are going to want to produce a lot of babies if most of them die young. Spider reproduction is like the old axiom about voting: it should be done early and – if you aren't planning on feeding your kids your rotting corpse, as some spider moms do – often.

The animal parent that should be celebrated for her contribution to our understanding of life history in mammals, who are a bit more like us than spiders, is the one introduced at the beginning of this chapter who I would like to refer to as Pizza Rat Hero Mom, even though there is no evidence that the rat in question was a mother, let alone a heroic one. Eating a pizza twice your body weight, while an initially appealing challenge, is just not something the average human attempts. For a rat, the calories in that pizza (about 190) are more than triple what the average rat needs per day (60). If we imagine that this random subway pizza thief is a mother, however, her heroism starts to make more sense. Mammals, it has been noted, are difficult to build. All that fur really does cost, plus there are brains and guts that are a bit more complicated than your average insect. One of the reasons Pizza Rat has to go to such efforts is down to the basic energy mechanics of reproducing 'expensive' offspring. Life in the subway tunnels isn't always easy – and even though she might have a dozen or so pups at a time, around half of them won't make it out of the nest. Both Charlotte the barn spider and Pizza Rat have reproductive strategies where their energy is devoted to maximising the number of offspring, rather than investing time in raising the offspring. It makes sense for rats to breed like, well, rats, because they've got to balance that investment in growing babies against the high likelihood that they're going to take some losses. This naturally leads to the sort of parenting that in a human would see social services called, but for our Pizza Rat it makes sense. The rat strategy is to have frequent, short pregnancies resulting in multiple pups that are themselves ready to breed in less than two months. This strategy makes for an exponential number of rats.

Our Pizza Rat, unlike the spider, will stick around a while after birth because her babies still need some parenting, but it's in her interest to move on quickly and have as many litters as possible. Leaven out the energy availability in an environment and the amount of energy that can be stored up in a little body with the statistics on how likely the animal is to survive any given amount of time, and we find that small, high-mortality, high-birth-rate rats are a logical response to the pressures of their biology and environment. To get the most out of their baby rats, our furry friends want to be able to grow them into sexually mature rats in the quickest, most efficient way possible, because Pizza Rat's babies aren't done growing at the end of the pregnancy. They don't 'balloon' off into the sunset like the barn spiderlings, perfect miniature adults clutching strands of silk. They are still only half-cooked, and they're going to have to be supplemented after birth until they hit self-sufficiency by whatever energy the mother can carry back into her offspring, for instance through her milk after eating a giant piece of pizza out of the trash.

Unlike the spider, Pizza Rat has an extra chance to provision her babies. She has milk, and the time she spends nursing her pups is the extra investment in childhood all mammals get. It also means she has to spend way more time dealing with her needy offspring than a spider does. This compromise, between how much the mother can invest in pregnancy versus how big the babies need to grow to survive, is why rat pups are born looking like they're only half done, or, in scientific terms, 'altricial'. Altricial species are born dependent – and altricial mammals are born with their eyes shut and lacking their fur coats. Usually they have to be stashed somewhere – nests, tree trunks, boarding schools – until they can survive on their own. The opposite of altricial species are 'precocial' – think of the charming knock-kneed neonatal sprint of the baby giraffe or the instant perfection of waggle-legged spiderlings zipping about, right after eating their mother.*

There is something else about the early knee-users versus the furless wonders. Imagine them, if you can, side by side, Pizza Rat with her babies and One Ton Giraffe Mum with hers. It's really hard, because rats are small and giraffes are enormous, and therein lies another almost-rule in the animal kingdom: precocial, ready-to-run animals tend to be bigger than altricial, helpless ones. And yet, here our species is, quite a bit bigger than a rat, still having useless babies. So, what gives? What is the benefit of being born kitten-weak and helpless? What is the downside to being born to run? The answer is in the timing.

For our baby rats, that extra period of dependency helps mama get them the energy she needs to feed them – she can't store it in her tiny little body, so she needs to parcel it out over a longer time and that's why they get stuck in the nest so long. While Pizza Rat is New-York-City-subway-rat big, maybe 300g, she is growing 60g-worth of babies, which is a fifth of her own weight. Now this is spread out over a bunch of babies at about 5g per pup – instead of having one 60g baby, she's hedging her bets over five to a dozen, but there's still just no way she can fund any greater investment in each pregnancy than that 60g – her body needs to function too. In the end, her babies will be born

---

* For fans of children's literature, barn spiders like Charlotte are *not* known matriphages (animals that cannibalise their mothers).

small, but there will be more of them and they will grow relatively quickly.

The baby giraffe, by contrast, will be born weighing 70-odd kg, though that's not actually saying that much when your mother is, quite literally, the size of a house. With mama weighing in at an actual ton (a little bit more actually, maybe 1,350kg) and just under two storeys tall (say, 5m, give or take), a giraffe's baby at 2m high is actually already 40–50 per cent of its adult height and 20 per cent of its weight. The baby rat is less than 5 per cent of its mother's weight. Even though both animals deliver babies totalling 20% of their weight, Pizza rat's multiple surviving kids are going to grow up to be multiple rat adults, whereas that single giraffe baby will only ever grow up to be a single adult. For Pizza Rat, living in the dangerous world of the New York City subway, hedged bets and minimal investment are the way forward. For One Ton Giraffe Mum, however, survival on the savannah means moving, so she will put the effort in during pregnancy so her baby is ready to run.

Back to the vexed issue of whether size really does matter. If we were to look at mammals in particular, as a general rule of thumb smaller animals (like rats) have 'fast' life histories, whereas larger animals (like giraffes) have 'slow' life histories. If we think about where our baby rats start out (5g) and where they've got to get to in around six months (around 300g for a girl, more for a boy), that's some growth trajectory. By contrast, it takes 15 months for One Ton Mum just to gestate a great big bouncing ball of unlikely cervical vertebrae and knee joints, and she will spend a further 16 months nursing that baby, and adulthood won't come for years yet. The baby rat is on the fast track and that baby giraffe is definitely on the slow, with an incredible absolute gulf between them in terms of how much energy is actually being invested in each. However, Pizza Rat has to provide for a much sharper increase than our slow-and-steady giraffe. Not only does she need to support exponential growth, she needs to do it in way less time than a 'slow' living animal. No wonder she's turned to pizza.

Given the proportionately smaller amount of growing a baby giraffe has to do to reach adult size – the critical evolutionary point when it can start producing new giraffes on its own – the 'born big' strategy seems like a clear winner. Thinking over the long term, the idea that organisms prefer to go big is called 'Cope's rule' and suggests that animals were all evolutionarily primed to try to get bigger. Dinosaurs got bigger over time, the rule says, as do mammals, molluscs and all sorts of things. You may have noticed, however, that this is not exactly the way things have worked out. Your local dinosaurs – pigeons, crows, sparrows – are actually quite small these days, and feather-light. The problem with Cope's rule is that it actually takes longer to grow a bigger animal. There's a pure physical constraint on how fast an animal can put on mass, meaning that even though the baby rats have further to go in terms of achieving multiples of their own mass, they're going to be able to do it faster than a larger animal because converting energy into 300g of rat is a heck of a lot easier than converting it into an actual ton of giraffe.[*]

Celebrity biologist Stephen Jay Gould reckoned the entire idea of an evolution designed towards supersizing comes more from us humans having a fondness for the 'bigger is better' school of thought than actual science, making Cope's rule just another example of how we attribute our own moral arc to the fossil record. In fact, all of the so called 'rules' we have about growth and

---

[*] To understand the difference in mass, a baby giraffe is approximately 14,000 baby rats.

development are problematic. It would be comforting to provide a slick formula to describe the relationships that govern growth and development in animal species, but the fact is, we still don't have an absolutely clear idea of what trade-offs, exactly, we are dealing with. The computational revolution of the late 1970s and early 1980s inspired a number of trends, among them an alarming fondness in academia for restating fairly simple arguments as equations of such baffling complexity that they physically defy reading.* However, just because you have an equation or a rule for something doesn't mean it's right – otherwise turkeys would be the size of freight trains and we would have to seriously rethink holiday dinners.

If life was as simple as a series of eponymous rules, then we could just draw a pretty equation with spiders down one end and giraffes on the other and triangulate where we ought to be based on our size at birth and the size we plan on getting to. A giraffe has a nice long pregnancy, and is born ready to run at about 20 per cent of its eventual adult weight. The rat pup, by contrast, is born quick and emerges helpless at a mere 5 per cent of its adult weight. If we humans start out around 3.5kg and are headed towards 60kg (ish – we're a bit big ourselves these days) then we're born at 6 per cent of our adult weight – much less than a slow-living, precocial animal like a giraffe and only a little better than a rat. This happens despite the fact that we have much longer, slower lives than that giraffe, so we know there must be something else in the equation, making our human life histories the odd little ducks that they are.

The difference between them is in the timing of when the investment in the next generation is made: when that

---

* Early scientific formula of how babies end up in the state that they do are an alphabet soup strained through parentheses and superscript.

energy is transferred from one generation to the next. We've got to think about how much energy the mother can store, how fast the foetus develops, whether the baby is going to be parked in a nest with its mother looking after it for ages or away gallivanting straight off – there are so many variables to consider that predicting a cross-species pattern becomes very difficult. There are however real biological phenomena that correspond to how you get animal babies born the way they are: the rate of growth in the womb, pregnancy length, how many babies at a time, the mother's size and how dependent or independent that baby needs to be. All of these factors make up the critical amount of investment – and timing of that investment – that it takes to get that baby to sexual maturity and a shot at winning evolution. All the animals that have ever walked, flown, parachuted on gossamer threads or carried a pizza have come up with life-history strategies that balance the risks of living fast and living slow, and primates – funny little big-eyed guys like tarsiers, lemurs, monkeys and apes – which is what we are, no matter how much we try to downplay it – are no different. The thing that changes is where, and how, the investment in the next generation is made. We call these stages of investment 'life history': the waymarkers on life's path of birth, growth, reproduction and death.

What we want to understand are the evolutionary pressures that shaped our life history strategies: how long we spend on each stage and whether our overall strategy is shotgun or sedate. This is complicated by the fact that there are so many potential evolutionary pressures out there and animals have wildly different reactions to them.* If we

---

* Take the exploration of space as an example – hardy, microscopic tardigrades react by doing absolutely nothing and humans send up cars with Bowie songs on the radio.

want to understand how human childhood came about, we need to put it in a line-up with the could-have-beens of animals with similar biological and environmental circumstances. We need to know what our options were, and for that we look to the monkeys, apes and lemurs cluttering up our family tree. If different life histories are the stories of different strategies of investment, we want to know where our species banked its cash.

There are dramatic differences in the life histories of the adorable mouse lemur, for instance, and the sober old orangutan. Being born a mouse lemur does not necessarily give you the best shot at life. About half of baby mouse lemurs will not grow up to reproduce and those that do rarely live to see a fourth year out in the wild. And there is a reason orangutan – 'forest person' (*uraŋ hutan*) in Malay – is so commonly misinterpreted as 'old man of the forest'; they can live to 60 and even a little beyond. A few years ago, forest person Puan, Perth Zoo's celebrated Sumatran orangutan, passed away at the age of 62, having produced 11 baby orangs. That's a solid achievement, given that most female orangutans only reach sexual maturity at around 15 years old and, in the wild, tend to space their infants eight to ten years apart. The mouse lemur, by contrast, might only have time to get out a handful of reproductive events in her short lifetime; even if these are the twin births that are the most common for her species, that still gives us a maximum potential of somewhere between four to, say, 12 baby mouse lemurs – and as we already know, half of them are for the chop.

Eleven offspring against, say, four seems like a fairly clear win for the slow-maturing orangutan. However, that slow growth is also a liability. Those 11 babies would have been more like four if Puan had been in the wild – assuming there was a wild left for her to be in. And 60 years to produce four of your species equals a lot fewer orangutans than the exponential number of mouse lemurs that might be produced in the same time – anthropologist Paul Harvey worked out the potential difference as 10 million mouse lemurs produced in the space of one (similarly slow-breeding) gorilla generation. Currently, both the mouse lemur and the orangutan are facing a very uncertain future. Habitat destruction is threating the island of Madagascar, land of the chirpy little lemurs, and deforestation in Indonesia – much of it driven by our need to have more cheap palm oil to grease up our processed foods – is literally burning the orangutans out of their homes. Mouse lemurs, long accustomed to a high mortality rate, are living fast, which is associated with living in an unstable environment that has a tendency to pick off youngsters. Even adapted to living fast, however, all the lemur species of Madagascar are in danger of becoming *lemures* (ghosts, to a Roman) due to environmental destruction. If that's the situation for the mouse lemurs, you can easily see the peril of the slow-living orang, which is following a life-history path adapted to a stable environment where years of investment in a single baby are likely to pay off. For animals that have taken hundreds of generations to reach an equilibrium with their environment, sudden change in the environments they have adapted to is worse than a disaster – it's a death sentence.

Where do humans sit on this scale, and why? How do we put ourselves into the broad spectrum of fast and slow lives, and what evolutionary pressures put us there? We are not as altricial, or helpless, as a baby rat at birth, but we are certainly not ready to roam the savannah. It is probably for the best we don't emerge fully formed in our hundreds to consume our mothers, because, like rats, puppies, sparrows and all sorts of other creatures, we need them. The investments we make in our offspring are a crucial pillar of our reproductive strategy. We have balanced biology against environment to come out with a very specific life-history pattern, one that we will draw out in the next few chapters as we compare our lives to the animals most like us. That life history explains how big our babies are, why we have more single births than twins, why we have babies with open eyes but useless motor skills and why it takes such a tediously long time to grow them. It also underlies the 'how often' of human reproduction, and the start and finish line of our reproductive lives and why, unlike so many other animals, we live beyond it. Critically, for the purposes of the discussion you and I are about to have, it also gives us a way to talk about the choices we make in how we invest in our children – how we grow our new humans.

## CHAPTER THREE

# Two Little Monkeys Jumping on the Bed: Making More Monkeys

Two little monkeys jumping on the bed,
One fell off and bumped his head,
Mama called the doctor and the doctor said,
'No more monkeys jumping on the bed!'

Where do children come from? Humans have come up with a remarkable number of euphemisms, metaphors and outright lies to describe our process of reproduction. You may be told, while still a child yourself, that babies are brought by storks[*] or come from peaches or are found under cabbages. This is patently nonsense, but the question Talking Heads asked of us all is important if we hope to understand how we ended up with the childhoods we have – how did I get here? As we saw with Pizza Rat, children and their childhoods take different shapes depending on a host of biological factors like size and growth trajectory. There is another dimension to the factors that decide what kind of life history we are going to live out, however, something that carries us beyond the simplistic 'be big, grow big' nature of much of the animal kingdom: our biology is mediated – supported and supplanted – through our behaviour. Before we even get to childhood, there has to be a series of behaviours that get us there, and number one has got to be reproduction. It's a

---

[*] There is, in fact, a strong correlation ($p < 0.008$) between stork population numbers and human population numbers, which is why you should be extremely wary of p values. And also, possibly, storks.

simple biological process,* but the way we do it is part and parcel of what shapes our childhoods.

If bees, birds and the more academically minded sort of fleas do it, why should we look to reproduction for information about the evolution of human society? It turns out even the process of making new humans is something we do through the lens of our socially learned culture. We've discussed the different overall strategies available to life on earth in the previous chapter, in terms of slow-and-steady versus all-your-eggs-in-one-basket types of life history. Humans fairly clearly have a slow life history, where we reproduce expensive babies in small numbers – more on that later – and then have to raise them. But how, you ask scientifically, because this is a book about human evolution and not *Cosmopolitan* magazine; how do you decide to make a human baby with its attendant slow life-history needs in the first place? While it may seem that there is a lot more that goes into arranging a human baby than any of our critter counterparts,† many of the basics hold true from birds to bees to you. Someone's got a big gamete (egg!) and someone's got a small one (sperm!); somehow they've got to balance their needs against those gametes and make it work.

Stretched analogies about bee social organisation and modern work practices aside, there are several aspects of our respective social organisations that make our reproductive patterns distinct from our bumbling friends. We are mammals, for one thing, and as such tend not to isolate the job of producing new members of the species into the personage of a solitary female while delegating

---

* Ha. Ha. Ha.
† I have yet to observe a cockatiel attempting to game Tinder, for instance, although I do not rule it out. Very vain birds, cockatiels.

large swathes of the population to the position of 'sexless drone'.* There are, of course, a vast array of possibilities for reproducing one's species, ranging from parthenogenesis (virgin birth) in certain types of sharks and lizards to reproducing sexually and ending up producing not just offspring, but scientist offspring capable of stripping genetic material from an egg and making a clone. What humans generally go in for – for now – is a little less complicated than Dolly the cloned sheep. It is actually rather run-of-the-mill sexual reproduction and fits well within the broader range of primate behaviour. Though that range is broad – very, *very* broad.† When we talk about human evolution, it is important to remember that what we consider 'normal' is defined by our culture, rather than biology. Sex is social and reproducing a species does tend to come down to sex. So, let's look at the options we have for society: the rules of how to reproduce.

There is a tendency, among people who have mostly been exposed to a specific narrative about human life, to believe very firmly that human reproduction centres on two individuals of different, absolutely dichotomous, biological sex. This is a narrative humans have told themselves in a number of different times and places – from thousand-year-old stories featuring gardens, snakes and apples, to present-day tax forms with a box to tick if you fit within a certain legal definition of pair-bonding. The 'nuclear family' as an ideal reproductive unit for the human species is, as a phrase, only about a hundred years old and until recently was a grouping assumed to have been more

---

* Yes, yes, down in back there, thank you, the joke has been made.
† Not as broad as penguins. The Victorians were so horrified by the homosexuality and necrophilia they observed in Antarctic penguins they censored their expedition notes.

or less invented by the Industrial Revolution. Many social historians have argued that prior to the growth of the modern economy most of humanity would have lived in combinations of generations and relationships – groups of siblings with their wives, for instance, or the still-popular grandma-at-home scenario. However, we now know that there are examples of all sorts of living arrangements within the long history of our species, so this begs the question – is whatever is the norm *right now* the thing we evolved to do? Were we born to be silverbacks (and/or have non-speaking roles in the harem) or did evolution push us into pairs? Is the nuclear family the evolutionary endpoint of all humanity or do we need our larger kinship groups in order to survive in this world? What are the larger implications of our social system when it comes to producing children and determining the nature of our childhoods?

How do we make sense of human reproductive strategies – assuming we can figure out what they are – in light of the evolution of the many mating strategies available? First, we can examine the biology of the thing. The biology of human reproduction is by definition asymmetric, lop-sided. Sexual reproduction in our species asks a lot of the female, taking up considerable time and effort. Males, however ... well. This discrepancy in effort is what makes different strategies of reproduction more or less successful for different parents. Females are only ever able to have a certain number of children, dictated by the length of their reproductive years divided by the length of time they are pregnant and then nursing a baby, and of course whatever other factors come into play that keep them healthy enough to get and stay pregnant. In a horrific factory-farm style scenario that even dystopian visionary Margaret Atwood would shudder to consider, for human biology that would be a maximum of one baby a year, for a potential 30 years of baby-making. That these numbers are never reached in reality is due to (hopefully) some level of female agency and (depressingly) that this level of child-bearing is pretty likely to kill you. Males, however, are capable of having a near infinite number of offspring, providing they find the opportunity. That we are not living in the world of *The Handmaid's Tale* just yet is due to the patterns of behaviour that have determined that actually, that's not a super ideal way to make successful batches of new people. But how did we get here? The clues are in the endless competition of our mismatched gametes: who invests when and how much.

The biggest investment is in, well, getting big. It turns out that size matters – sometimes. Sexual dimorphism – differences in size or shape between males and females in a species – is most pronounced in those primates whose males do a fair bit of competing with other males to win over the ladies. This very mammalian trait of big males is a bit of a reversal of the pattern in much of the animal kingdom, where females usually get larger than males, and is argued to be the result of sexual selection due to male competition in mammals. Why only mammal males should compete over mates and get bigger when competition is a strategy employed across the animal kingdom has always confused me. Perhaps it is not males that are big, for example, but females that are small, because they had to get on with turning over their energy to the next generation. However, most overarching theories insist that sexual selection, in the form of competitive male mating strategies, are why mammal males get bigger than females. Perhaps the problem is with what we choose to measure – it's certainly much easier to identify baboons screaming at each other on the savannah than to quantify the rather tedious work of gestating and nursing a baboon baby,* however that doesn't mean one has an effect on reproductive fitness and the other doesn't.

Let's go back to those bad-boy baboons for a minute though – particularly their enormous fangs. Frequently bared to fight off rivals, a male baboon's canine teeth can be 400 per cent larger than those of a female, because she doesn't need to fight anyone very often, she just needs to eat grass and the occasional small insect.† Humans, by contrast, only have a size difference of about 7 per cent in our fangs – and even that is pretty negligible once you get a broad enough range of people.‡ We're not a terribly dimorphic species in any part of our bodies, no matter what our aesthetic industries try to tell you. But some primates are – and it can cost them. 'Weaponised males' like baboons pick up injuries from fighting with those teeth, as do mandrills, macaques, gorillas and other big-boy species. In addition to the extra energy required to grow bigger teeth, or a bright red nose, or just bigger overall, those males still have to find enough energy for that mate competition. There is evidence that mating seasons sends some primate males into a state of real physiological stress – and it certainly can't help your chances of surviving a leopard attack if you only have your eyes out for the ladies. A tree-dwelling

---

* The observation that courting male baboons 'often look exhausted' is unlikely to be met with much sympathy from any primate that has ever been pregnant or nursing.
† And if you can see a baboon's teeth, you had better think about leaving. Quickly. Humans are the only apes that mean something nice by 'smiling' – and even then only some of the time.
‡ I have a longstanding and utterly non-scientifically tested theory that excessive canine tooth size is one of the reasons behind the sexual appeal of vampires and David Bowie. That and dress sense, obviously.

monkey exhausted from mating, attempting to mate, or guarding his mate is dangerously more likely to be an abruptly non-tree-dwelling monkey.

Physical investment in reproduction is not, of course, limited to teeth. Unsurprisingly, the genital organs themselves are subject to some adaptive pressure, to the point where distinct patterns of genital size can be observed in primates with different mating patterns. Chimps, for example, have enormous testes relative to their body size – an average of about 118g of scrotum to 44kg of chimpanzee, or 0.2 per cent of their body weight. This, it has been argued, is because chimps need to produce a lot of sperm in order for their own genetic material to have a fighting chance, because chimpanzee females usually mate with a large number of males and there is likely to be quite a lot of competing sperm present. For chimps, devoting developmental energy to growing more sperm-producing tissue is a sound strategy – as it is for baboons, macaques and other primates where multiple males might mate with a given female. Mouse lemurs are the primate champions in this category – the tiny squirrel-sized prosimians weigh all of 280g but their testes are around 15cm$^2$ in volume – the equivalent of a human male having testes the size of grapefruits. A 'scramble competition' mating strategy leads to sperm competition, and that means big testes for males and sperm that is either better or worse at sticking around. In females, it might also mean going into heat (oestrus) only at a specific time, narrowing the window for reproduction and increasing the pressure on those males. It might also encourage the formation of what's called a copulatory plug, which is a female strategy for physically sealing some sperm in – and any additional deposits out.

What about mating systems with less scramble? For mating systems where there is only a single male around, like the gorilla with his harem, the ratio of testes to body size is much smaller: 170kg of silverback comes attached to a measly 30g of testicle (0.002 per cent of his total weight). Monogamy, it seems, also breeds tiny testicles. The underloved gibbons – who are apes but not 'great' ape enough to hang with the chimps, orangs or gorillas we find so fascinating – are nice little primates that form pair bonds and carry about 5.5g of testes for each 5.5kg of gibbon, 0.1 per cent of their body weight. And humans? Well, we're not on the large side – a measly 40g of testes for an average of 65kg of person, 0.06 per cent of our body weight. If you plot our position on the arc of primate mating strategies based solely on testes size, we're in there between one-male mating systems and pair-bonded monogamy.

Testes size is not the only tell-tale sign of a mating system. Animal genitalia by necessity adapts to the task of delivering and receiving procreative *matériel* depending on the reproductive environment. The vagina and phallus[*] also respond to adaptive pressure, with shape and size of both organs changing to maximise the potential of having offspring (for males) or worthwhile offspring (for females). We know relatively more about the phallus, hopefully for the reason that it is easier to observe and not because it is a subject of almost pathological interest to a certain type of anthropoid primate who goes around conducting anatomical surveys of apes and monkeys, but it should be remembered that for every new penis flange that evolves, the female of the species will have to have a response – and one that suits *her* evolutionary priorities. The primate phallus, or 'internal courtship device', doesn't quite achieve

---

[*] Or penis. Or, if you're particularly keen to get as much scientific terminology as possible between yourself and the thing, the 'intromittent organ'.

the variation seen across the animal kingdom – there are no corkscrews, for instance, like the surprisingly outré mallard duck* – but there are differences in length and shape that relate to mating strategy. In primates, the more complicated the penis – the addition of penile spikes,† elaborate distal ends, an 'os baculum' (penis bone) and extra length – the more mate competition there is likely to be *inside* the female; the elaborate penis is an attempt to ensure that a given animal's sperm makes it up to an egg, rather than that of any other males that have already copulated with her. This is called 'cryptic female choice', which is a sort of weird conceptualisation (literally) of the female reproductive tract as a black box into which many sperm go in and only one winner comes out.

It has been posited, in contexts both academic and non, that humans have quite large phalluses for their body size in comparison with the rest of the primates. As discussed above, a bigger, more complicated penis would suggest that competition for mates was an important factor in our evolution. However, as much as humans might like to believe otherwise, our penises are nothing particularly special. We are in the top 15 or so for length and do very well on girth, but overall the human penis is perhaps one of the most basic primate penises available. We lack the spines, complicated end, baculum and length of a devoted polygynous primate. Human males aren't even as good at producing ejaculatory fluid or sperm as those animals that depend heavily on sperm to do their competing for them. Human sperm is too viscous to serve as impediment to

---

* Special mention for Jules Howard's book *Sex on Earth* here, and certain educational moments during his live talks that cannot be unexperienced.
† Yes, you read that correctly.

competitors. Chimps, on the other hand, have quite sticky semen that may serve to block out other competitor material.

Frequency of copulation is also not a human strong suit, with lower average numbers of copulations per hour than the more competitive primate species and decreasing sperm levels (down 85 per cent after two to three ejaculations), while that chimpanzee is ready to go again after about 5 minutes. Even the time it takes to copulate is subject to evolutionary pressure. Not only is longer copulation, or copulation where the two parties are locked together, thought to help keep out competing male genetic material, but the environment may be a contributing factor as well. Arboreal species take rather more time about things than terrestrial ones that might be easy targets while, cough, distracted – galagos are at it a full hour, while orangs last about 14 minutes. The Kinsey report on sexual behaviour, meanwhile, grants human males an average of 2 minutes. None of this speaks to a recent evolutionary history of vastly competitive multi-male mating systems.[*]

Female agency in primate mating is a rather dismal topic, whether you're reading *Folia Primatologica* or the gossip pages. A lot of this, anthropologist Catherine Drea explains in a recent review, is because no one from Darwin on down could really imagine what females in charge of their own sex lives would look like. We developed our idea of what sexual selection is, in functional biological terms, by watching fruit flies go at it in the 1940s. Fruit fly males have pretty different rates of reproductive success, meaning that any extra boost to reproductive potential really increases the chances of your genetic material getting into

---

[*] Or much time in trees.

the next generation. This lies behind the idea that male investment in impressing the ladies is the most critical evolutionary behaviour, and became normalised and formalised by the rather complicated biologist Robert Trivers* for primates in terms of who has to work hardest to make a baby. Cheap sperm versus expensive eggs – and pregnancy and the rest of it – are formulated as a constant conflict, with males trying to mate all the time and females trying to be choosy. Well, what is fine for fruit flies is not fine for primate females. We have to move away from the idea that the only way mates get chosen is through some monkey mixed martial arts ultimate mating competition.

Primates have complicated social lives and this comes out in female reproductive strategies – when researchers can be bothered to look. Female primates don't always have sex just for reproduction – bonobos, for instance, rather famously use sex to socialise. The extent to which sex was a social act, rather than a hardwired biological tool for evolution, came as something of a shock to early researchers, who suggested that the first homosexual behaviour observed in two younger male bonobos only happened

---

* Robert Trivers has had an interesting career, to say the least. Having made an immensely impactful study of sexual selection in primates, he went on to the highest echelons of research and there attracted the attention of the notorious child molester Jeffrey Epstein. While repudiating Epstein's predation of underage girls, he did manage to make the statement, 'By the time they're 14 or 15, they're like grown women were 60 years ago, so I don't see these acts as so heinous,' which is nonsense, and insidious, illegal, sexually exploitative nonsense at that. While admitting *saying* this was a terrible mistake, the dude still said it, and I'm just going to leave that there so we can all think about the environment that top-flight primate sexuality research has occurred in.

because they were locked up in a zoo.* Primed to expect a dominant male and submissive females, or possibly a male–female pair, the fact that bonobos happily get off with either sex with a variety of appendages and in a variety of ways forced primatology – and the rest of the world – to rethink what is 'natural' about sex. Bonobo sex isn't always about reproduction – they use sex to calm down after fighting, to make and keep friends, and to stave off boredom. In other words, bonobos, like many other primates, have 'communicative' sex – social sex, which cements alliances, arranges hierarchies and generally performs all sorts of functions that don't involve conceiving a baby.

Some female primates – human women, for example– have gone one step further in the 'sex for other reasons' category and developed 'concealed ovulation', which means no one (including the primate herself) knows quite when she's fertile. This untethers sex from reproduction even farther. In many species oestrus is announced loud and clear through a series of audio or visual signals. A female cat, for instance, is in heat for a limited period of time and needs to attract suitors† at exactly the right point because there is really no reason for male–female cat interaction other than the act of mating. For most primate species 'the right time' is fairly clear from highly visible swelling or colour changes around the genitals, and in others there are much more subtle changes like faint reddening in facial skin tone, or fidgety little changes such as social behaviours and smells. The bright-red butts of several types of

---

* Fascinating Freudian question there in what life experiences led people to think zoos lead to homosexuality.
† If you have ever heard a female cat in heat, you will understand why English contains the unique word 'caterwaul' just to describe it.

monkeys – from macaques to baboons to chimpanzees – are an example of the fairly obvious visual signals of fertility common in other primates. These changes accompany the hormonal cycles that determine when ovulation, or the window of fertility, occurs and in most primates the idea seems to be that advertising the moment of peak fertility is the way to get yourself a baby.

Ovulating human women display very subtle changes that you'd be hard pressed to pick out, compared with, say, an olive baboon whose swollen genitals can be as much as 14 per cent of her body weight.* Why we decided to hide when we are fertile occasions a great deal of argument in academic circles. Some say that keeping ovulation a mystery is a way to make paternity equally mysterious and maybe save your offspring from any infanticidal males who are looking to clean house, though there are still problems with the whole idea of infanticide being a concept that primates understand. At least one argument says that human females have evolved concealed ovulation so it appears they are *always* fertile, with our noticeably fat and hairless butts mimicking the sexual swellings of other primates while we are in our habitual bipedal posture.† Opposing this are the folks who say we stopped being obvious about our fertility when we started standing up, because standing up hides our genitals. Welcome to the world of evolutionary

---

* Unless, of course, the human female is willing to invest in a vaginal temperature reading every day. And given the booming industry of fertility-testing and tracking apps and products, apparently we're up for it – and rethinking the wisdom of hiding ovulation.
† Yet we mostly use our large butts for sitting, while the poor olive baboon who is using hers for sexual display finds sitting pretty uncomfortable when she's ovulating.

anthropology, where there is reasonable explanation for everything – and several unreasonable ones if you prefer.

The reality of the situation is that while there are only minor physical signs of ovulation in human females – changes in body temperature, flushed skin – there are more significant changes in mood and behaviour. Female behaviour might, rather unsurprisingly, be a big part of fertility strategy. According to research that examined how turned-on women have reported feeling towards their partners versus other males at different times in their cycles, it seems that women don't fancy their partners as much when they are ovulating. Human women also don't limit sex to times they could get pregnant. One of the major questions with primates, actually, is why, exactly, females waste their time on non-reproductive sex. Other animals don't bother, and given the amount of energy and hassle involved, plus the risk of aggressive partners and other disasters, it is a legitimately questionable behaviour. However, by being sexually receptive throughout their cycle and concealing when they are actually fertile – and into other guys – they may give themselves a better chance of getting away with widening their offspring's gene pool, either by mating with several males, who will then all feel bad about killing any potential offspring or by cementing relationships that will benefit the mother. Promiscuous female sex may even use up all of the sperm available for making new babies, making sure other females are left without offspring while they hoard all of the valuable gametic resources to themselves.

Primate female mating strategies run the gamut then from giving no hint of ovulation to waving pink or swollen genitals about in direct competition with any other female also trying to advertise for a boyfriend, through to, possibly, sperm-hoarding. In addition to competitive advertising, females might band together to influence mating strategy. For instance, those free-loving female bonobos get together to support their favourite males, forming coalitions that make or break dominant status and mating success.

There are other avenues for female competition strategies as well and one of the most interesting is the different use of what is called a 'copulation call' – essentially the noise a female primate makes during sex. These noises can act as inspiration for her partner, with calls determining the time of ejaculation in some macaques, or they can function as general braggadocio, letting everyone else around know what a good time she's having. While that might sound like an odd motivation for a primate, female chimpanzees are louder when they're with a high-ranked male, someone who might be beneficial for both mother and potential baby to have on side. However, if other females are watching they keep shtum to a much greater extent, possibly to smooth over any potential for competition and female aggression. Bonobos, who mate – or at least engage in sexual play – with anyone, at pretty much any time, also change the noise they make depending on the social situation, being louder about some partners and some types of sex than others. To some extent, primate sex noises have become part of our socialisation repertoire, communicating far more than the fact of reproductive copulation. It's possible that the complicated sexual politics of our ancestors was the thing that made language necessary in the first place, although our anatomy seems intent on signalling that we are not on the competitive edge of the sexual spectrum. We've got all the hallmarks of a monogamous species – or do we?

## CHAPTER FOUR

# A Froggy Would A-Courting Go: How Weird is Monogamy?

Froggy went a-courtin' and he did ride, hmm, hmm,
Froggy went a-courtin' and he did ride,
A sword and a pistol by his side, hmm, hmm.

All animals have established strategies for deciding how the very first investment in the next generation should be made. These get solidified into rules about reproduction, or 'mating systems', which structure animal behaviour and society. Primate social structure is made up of a dizzying combination of group sizes, ratio of males to females, long-term or short-term associations; this means we can end up either with a silverback dad, or piggyback-ride dad, or even us with our Adams and our Eves and our tax forms. Anthropologists use the term 'social organisation' to define the interactions that make up the internal organisation of primate life, things like who you groom, and who you live, mate or otherwise have to deal with. Primate 'social structure' exists as a long spectrum from hermit straight through to party animal. Social organisation – who your friends are – is just as varied. The very basic structures of social life and our patterns of reproduction are inexorably linked, locked together in a feedback loop.

So, what are the options for structuring a rag-tag society of primates? There are a truly impressive number of potential arrangements of primate social structure. You might get adult males as solitary units, or seasonal solitary

units, or solitary units while they're a certain age; you
might get related females with unrelated males, unrelated
females with related males, or either of those with a single
breeding partner of the opposite sex. A group of females
with a single male is common, as are groups of multiple
males and multiple females; smaller groups within the
larger might form based on sibling bonds, female soli-
darity or age groups. You might live life as a pair of
gibbons, barely tolerating the neighbours, or you might
reduce your domestic trouble by half again and live like
the orangutan, puttering about the Bornean rainforest,
possibly occasionally with a kid in tow, but otherwise
gloriously, happily alone. We also have mandrills, which
are extraordinarily handsome monkeys of the Old World
variety that live together in sort of meta-groups of adult
females and children – with the occasional intersection of
multiple or individual roving males – creating groups that
are so large (around 700 individuals) primatologists have
simply given up and called them 'hordes'. Far less common
is a third primate social structure: the pair. Only about
10–15 per cent of primates live in such reduced groups.

When it comes to social organisation of primate life,
there are a whole new set of variables. Who you forage
with may not be who you sleep with, and who you sleep
with may not be who you reproduce with. Infants may be
cared for by non-parent males, or females, or they might
be killed by them; they might live or die on the status of
their parents or they might be more of a group project.
Offspring might reach adulthood and live with their
mother, or go searching for a new group; solitaires and
pairs might criss-cross the territories of their sisters or
other relations or spend half their time sneaking off to find
novel mates. It's a wild world, after all. But what do these
variable solutions for primate living mean? How does a

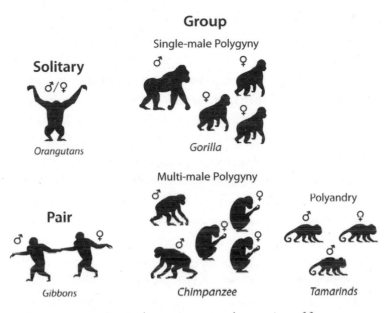

Figure 4.1. *Primate social groupings can take a variety of forms.*

certain group size or tendency to live with your mother tell us about what environmental or adaptive pressures shape the way we mate? As you might imagine, the organisation of a given group has an effect on the mating system – and the mating system determines which babies exist – and who gets to be a parent. Even though primates have the burden of reproducing sexually, using two biological sexes, what little room for diversity and flair has been allowed has been enthusiastically seized upon. There are a number of polygynous (multiple-partner) mating systems that involve either a high-status male or high-status female, bossing around a subordinate harem – that marmoset matriarch, for instance, if she picks up more than one male, stops being monogamous and becomes polyandrous. Polyandry describes a mating system of one female, multiple males; polygamy, one male, multiple females. You've heard of polygamy because it's still quite

popular in some *sapiens* circles,* but there's actually several
ways these strategies can play out in primates. In the aptly
named 'scramble polygyny', solitary males madly scurry
around looking for eligible females, a whirlwind of
one-night (or 10-minute) stands. Scrambling around is
still better than the more competitive female defence or
mate-guarding strategies, where males will do their
damnedest to keep other males from copulating with
females they're guarding. Then there's sneaky copulation,
again by males sneaking behind the backs of more socially
dominant males. As we saw previously, directly competitive
strategies are dangerous and hugely costly for males – for
one thing, they need to not get injured and die and for
another they've got to grow big enough to win.

In primate society, not everyone has an equal chance at
parenthood. Mating systems range from male monopolies
to female ones and, like any monopoly, there are going to
be losers as well as winners. The gorilla silverback is
perhaps the best-known example of a chest-thumping
male monopolising all the ladies and, charmingly, has
long been taken as a metaphor for human male–male
competition, despite the fact that of all the living apes the
one we are possibly least like is the ponderous vegetarian.†
Having mixed-sex groups is no straight line to glorious
egalitarian mating practices either. Dominant animals
may still throw their weight around, denying less
impressive neighbours a chance to mate. Even when pairs
are a more common occurrence, there's still no guarantee

---

* Even then, it's really not necessarily what you think from watching
episodes of *Sister Wives*. Some supernumerary wives are obtained to
relieve the labour of senior wives. And there's a lot more multiple-
husbanding going on than you think.
† All of my neighbourhood in north London excepted.

someone isn't being oppressed right out of the gene pool. Take, for instance, the deceptively cute and harmless-looking marmoset, commonly held up as a furry-eared exemplar of mating for life. Life in monogamous paradise is not quite what it seems from the outside for our tiny-faced friends, however.* Even through there might be many males and females living together, there is likely to be only one mating pair in a group because a single female boss marmoset will suppress the fertility of all other rival females, using an unpleasant combination of glandular secretions and violence to maintain her breeding dominance. Adam and Eve it is not.

Because primatology was for a very long time a boy's club, most of the mating strategies we think of feature males in the leading role, either competing against other males or 'sneaking' around them to get with girls that are otherwise out of their league. You can see how easy it is to anthropomorphise these types of strategies, which really drives home how odd it is that we rarely hear about how female-led strategies such as concealed conception or mate choice (or secreting reproductive-suppressing pheromones out of your behind) may have evolved, when at least some of these strategies are very obviously employed by humans. But in the heady early days of anthropology, it seemed absolutely certain that we would identify the evolutionary basis of everything – our social structure, our social organisation and our mating pattern – by making careful study of living primates as 'how we used to do it' and calling 'evolution' anything we did differently. Everything could be explained simply and functionally.

---

* Marmosets look like someone stuck a tiny normal monkey in an Ewok costume, then blow-dried it.

Our more solitary cousins, like the fabulously bug-eyed tarsier, are said to be avoiding causing a ruckus because they make their living creeping through the night, avoiding predators and searching for food. They can have their weird pair living. By contrast, our more gregarious primate relatives are thought to be avoiding being leapt upon by leopards with a 'safety in numbers' approach. Functional and environmental factors could explain every aspect of primate lives: resource-poor environments can only feed so many primate mouths, but more primate eyes looking for food might be a net win, and all of these factors further influence group sizes and the accommodations to social organisation that have to be made as a result. For the handsome mandrills, this means war.

While a wily primate would want company for protection – those mandrills kicking about on the edge of the savannah would be big cat kibble without their horde to protect them – there would need to be some sort of mating strategy to deal with the sheer number of monkeys running around. For human scientists who are culturally trained to see competition as natural and increasing group size as a green light for domination and hierarchical behaviour, it was clear that mandrills would have had to come up with a much more competitive mating system than some simple tarsier pair bond. And mandrills do, in fact, compete for mates. The development of what is fairly accurately called a scramble mating system means mandrill males fight each other for the right to mate with the females. In the just-so retelling of evolutionary development, these hyper-aggressive mandrill males wreak havoc on mandrill family life, so must be banished to the edges of society until needed for mating, and so on and so forth until you have come up with a weirdly creepy-male-sex-offender explanation for the entirety of mandrill society. And this is where we have to be careful. Mandrill male murderousness is impressive, much like their bright reddish-purple behinds. But it is equally possible to explain mandrill society as regulated by the most effective distribution of female caregiving and social relationships than as by having to keep mandrill males away. When we discuss mating systems and the societies that support them, we have to remember that there is more to mating than men – no matter what our cultural biases might say.

Discussion of male dominance in primate society is about where any humans reading might start to feel a sense of familiarity. There is a tremendous longing in certain sectors of humanity for the hypermasculinity of some primate societies, easily identified by repeated pseudo-scientific insistence on the relevance of 'alpha males' and 'silverbacks' to the organisation of human society.* There are evolutionary narratives that continue to pop up in the public imagination, presumably because they can be shoehorned into fitting with current western socio-cultural norms, which tends to lead to its own head-clutching crisis in evolutionary anthropological circles. For instance, a study that decides to test whether women really do prefer the colour pink – and therefore every stereotypical aspect of gender has an evolutionary basis – surveys a bunch of undergraduates in a UK university† and comes out with the answer that yes, yes they do. Even better, these conclusions are rendered

---

* Why humans should focus on violent male-against-male conflict in low-density groups of vegetarian foragers like gorillas as a framework for the vastly cooperative, hugely dense we-will-literally-eat-any thing-especially-if-its-fried societies of humans is perhaps something best left to the mental health professionals.

† The problem with asking anybody anything for science is that researchers never ask 'anybody' – they ask people in WEIRD – Western, educated, industrialised, rich and democratic – societies.

universal because the paper included Chinese students and therefore because China is not the UK any shared preferences had to be universal. The fact that they'd surveyed Chinese students who had presumably been exposed to the same cultural norms as their fellow students to the point that they were happily attending university in the UK seems not to have bothered the research designers one whit.

Anyhow, this kind of study – girls like pink! Men throw balls! Humans are into fried things! – is instantly awarded a deep evolutionary significance, because everything must be adaptive, somehow. There is even an entire academic discipline – evolutionary psychology – devoted to identifying the evolutionary principles behind our every action and choice. So now, a study reports that human females evolved to like pink, and the media runs with the story until we get to a point where it's being reported with weird post-hoc conclusions that try to explain the results in evolutionary terms; millions of years ago some ancestress was awarded better reproductive fitness by a mild preference for strawberries or some such thing. It does not matter that there are more poisonous red fruits in the world than tasty ones, or that what we think of as a strawberry was a hybrid of French and North American varieties invented in the seventeenth century. What matters is we have a reason for women liking pink and we firmly believe women like pink. Whether the vast amount of cultural baggage the study subjects carry with them includes very strongly gendered rules about the colour pink is not considered important. This is *science*.

Except, of course, it's not. It's a story. And it's only half of one at that. The more astute reader will have noticed that the entire discussion of mating systems features a lot of exciting choices, benefits and trade-offs for males. Mate guarding, male competition – none of these strategies have an active role for the female primate.* You can't actually get to primate babies without primate females, however, and what is good for the goose is unlikely to be equally good for the gander. Having long described primate mating patterns as a contest of competing male strategies,† a sudden, game-changing discovery was made in the late twentieth century – females.

Just like our old friend Pizza Rat, primate females also have evolutionary needs: to balance the energetics of gestation and nursing, to not get eaten by an eagle while lumbering around heavily pregnant, to have a baby that doesn't get marauded. Mating strategies become very complex when you have to describe the motivations of the females involved. Female baboons, it was observed, were not motivated to have constant swarms of males hanging around, pestering them for sex. Female marmosets were very much motivated by males who were willing to spend a full year carrying their stupidly heavy babies around. This is of evolutionary importance and not just an anthropomorphising swipe at lazy intellectual misogyny, because the female primate actually does have some agency in mating strategy. What she wants counts. This is one of the most critical things we have to understand about the process that produces human babies. There are a wide range of social behaviours that are being carried out by actors not necessarily with the same aims, but towards a generally agreed goal: successful offspring.

---

* This is true whether you describe success as the ability of primates to reproduce, or as the ability of primates to produce reproducible research outcomes in evolutionary anthropology.
† A tradition that the genre of 'hack your date' type advice columns, books and confidence tricksters is gleefully smearing right into the current century, under the guise of 'science'.

So what about us? What strategy got us here, with our concealed evolution, same-sized males and females, and social sex? We have all the hallmarks of a monogamous species – but what does that mean, really? And why does it matter for our childhoods? As humans, we have devoted immense televisual and print resources to applying this question in select cases,* and what we think of as monogamy a good divorce lawyer would demolish in a minute. When you actually tally up what primates are doing in the wild, it turns out that 'monogamy' is pretty ambiguous. It might mean having babies with just one partner or it might mean hanging out with one partner, but having babies with several partners on the down low. It might mean pair bonding for life or it might mean pair bonding until you find someone better. However you define it, it is in fact exceedingly rare in the animal kingdom – fewer than 10 per cent of animals opt for monogamy and even fewer if you leave out birds, 90 per cent of whom are dead keen on pairing up. Other mammals are generally monogamy averse, with only about 5 per cent of species having a monogamous mating system. However, about 15 per cent of primates, who, admittedly, are very weird, settle for 'The One'. The question is – why?

Monogamy in primates is almost exclusively observed on the small, adorable, lemur-laden branches of our family tree that are the farthest away from our own. Gibbons and siamangs, marmosets and tamarinds, those lemurs; while gibbons at least are apes, these are not our closest relatives by a long shot, but they are practitioners of the pair bond,

---

* One wonders if a primate freed of the bonds of the *National Enquirer* would be similarly freed of endless debates about which physical iteration of which multigenerational land-banking genetic conclave was sired by which celebrity.

much as we are. So, how did we get here? An array of theories for the evolution of pair bonds exists. One is that pair living results from females trying to recruit protection, either from males coming in and killing their infants, or from general male aggression. Another is that pairs result from males devoting themselves full time to making sure no one but they have a chance to mate with a female. Pair bonds might have developed to secure access to mating when females are scarce and prone to wander, or for females to secure a male defender against a high risk of predation.

One of the theories behind the evolution of pair bonds is that it allows a male animal to 'know' which offspring are his. Far be it for me to accuse early naturalists of having an obsession with paternity that is weirdly in tune with the preoccupations of a patriarchal system of capital transfer between generations and not the immediate evidence of the natural world, but this is a very odd thing to think early members of the primate family tree were actively thinking about.* Perhaps one of the most extreme suggestions has been that infanticide – or rather the threat of infanticide – is the most important driver of the evolution of social monogamy. An impressive recent survey by Kit Opie and colleagues studying hundreds of animal species does in fact identify infanticide as the key variable in driving monogamy. The thinking behind this is that infanticidal males kill children that are not their own, seeking to supplant them with their own offspring. In order to assure the paternity of a child belongs to a particular primate father, the female will seek to form a pair bond. Other factors, like the level of investment given by both parents to the baby and the

---

* Far, in this case, clearly meaning the linear distance to the footnotes, which is to say I damn well meant to say it.

need to secure access to women who can just walk away, are far less important. So let's look at infanticide.

In the mid 1970s, when eminent primatologist Sarah Hrdy was a doctoral student, she went to the slopes of Mount Abu in Rajasthan to observe the local langur monkeys for her thesis. She noted that male langurs would frequently stage hostile takeovers of groups, killing infants that were unlikely to be theirs. What might seem like an uncalled-for murderous rampage, however, Hrdy argued, was actually a perfectly adaptive murderous rampage. By ridding the group of infants of uncertain paternity, the marauding new males could make way for their own offspring. It wasn't only that there wouldn't be any competition from previous generations for resources – it was that by removing the need to breastfeed those infants, the females in the group would more rapidly return to being able to reproduce.

The evolutionary significance of infanticide has, at its heart, a rather fascinating supposition recently called out by the anthropologists Holly Dunsworth and Anne Buchanan in an essay for *Aeon*: that animals, from pea-brained mouse lemurs on up, understand the concept of paternity. Considering the huge mental ask of dealing with the conceptual complexity and long time frames involved in associating mating with birth, the authors wonder if it isn't a bit more parsimonious to argue that infanticide, rather than being a targeted evolutionary strategy, isn't just a bit more random. Perhaps males tend towards killing infants, but just don't kill infants clinging to females they have an affiliation with, without having worked out whether that baby is 'theirs' or not. Given that most human societies have had until recently an at best highly confused idea of conception,[*]

---

[*] And this confusion lives on in a deeply worrying number of elected politicians.

it is difficult to understand the mechanism by which selective infanticide operates.

A recent review of the quality of the evidence compares the anthropological obsession with monogamy – and the causes of monogamy – with that of a certain fictional FBI Special Agent Fox Mulder:

'If there were a high risk of infant-alien abduction, then they would be monogamous.

'They are monogamous.

'Therefore, there was a high risk of infant-alien abduction.'*

It's a fair point. We use monogamy to mean everything from Abrahamic marriage rules to the genetic make-up of a baby to two tamarins upping their levels of oxytocin by grooming each other.† Sometimes it's for life, sometimes it's for a season, sometimes males get replaced and sometimes females do. However we identify it, we demand this particular habit of pairing off to be 'because' of something – infanticide, baby care, how far a female wanders – and in many cases these may be incidental, consequential or otherwise just traits that happen to have co-occurred because we're not actually all that good at understanding which factors push and pull evolution in any animal, let alone complicated social ones. Evolution is difficult and we discover new factors – like women – all the time that force us to recalculate the models we were so proud of just years previously.

It turns out to be incredibly difficult to unpick the adaptive significance behind human behaviour because everything humans do is culturally mediated. So, we find

---

* Sorry, this is a traditional footnote. Fernandez-Duque *et al.* 2020.
† Tamarins, it must be said, have some of the best hair in the primate clade.

we have to be very careful with the questions we pose about what is 'adaptive' or what we have 'evolved' to do. History being what it is and humans very frequently being what they are, it took some effort to move our understanding of our evolutionary past beyond comparisons of gorilla harems and what one worries was a view of human society tinged by wish fulfilment. The just-so nature of primate social evolution (man wants sex, woman wants man and/or electrical appliances for the home)* was all so blindingly clear that it's actually a kind of miracle we ever stopped to question it at all. But miracles† do happen in science – and in the 1970s particularly they started to happen to anthropology. This is how we started to approach the more nuanced understanding of our evolutionary history we have today.

Humans, it seems, are rather bad at imagining how humans live. The popular imagination has over the years tried to fill in the gaps in our cultural knowledge with titillating supposition. It is hard not to suspect the same instinct that led to the grossly unpleasant fetishisation of some imagined 'primitive' state of human sexuality lies behind the fascination with polygamous cultures like the TV show *Sister Wives*. Human mating systems are, in fact, very circumspect. Even in social systems that allow multiple wives or multiple husbands (which do exist), the majority of human reproductive relationships fall into a pair-bond category – it's just that those pair bonds may not be lifetime-exclusive or mean what you expect them to mean. An interesting study among the Himba group from

---

* It is entirely possible, reading early literature on the mechanisms of evolution, to come to the conclusion that females weren't really fully involved with the process. Looking at you, Charlie D.

† Feminism. Or possibly just science. Very hard to tell sometimes.

Namibia makes the point that extra-pair paternity is highest when the female partner has less say in the marriage, suggesting that despite the evolution of a highly proscriptive, codified mating system in most human societies there are still multiple avenues through which sexual selection can act.

Even in one of the most monogamous primates, the lemur pairs roaming Madagascar, genetic testing reveals that it may be time for prosimians to get a reality TV show because more than 40 per cent of offspring from supposed monogamous pairs turn out to be someone else's. In human populations, the rate of extra-pair paternity is thought to vary from about 1 to 10 per cent; it's a little difficult to work out an exact number, as it turns out to be a pretty emotive subject when scientists start asking questions about it. Estimates of up to 30 per cent of children being 'passed off' to non-biological fathers as their own may actually suffer from what I'd call the 'Jerry Springer effect' – it is cases where there is existing uncertainty about paternity that tend to be tested, the underlying correlation may be with the potential for multiple candidates for fatherhood in the first place. And while, yes, extra-pair paternity does exist, it actually exists at a lower rate than in many other animals.

At first glance, monogamy does not sound like a particularly brilliant idea. It's a resource sink. For males that actually do parenting, hauling around a dependent is a terrible drain on resources, and perhaps the worst thing you could do if you were looking to trade up for a better mate. Dad is wasting valuable calories carrying that kid that he could be using to hone his mate-competition fang-fighting technique to win over a particularly fascinating female. Mama could be doing literally anything better than maintaining the social bond with dad – socialising with her friends and relatives, finding more food, judging fang-fights.* But neither partner in these pair bonds can take the time to engage in these aggressively selective pursuits because they have a baby that is completely and utterly useless. It needs them. Their potential reproductive efforts have been put on a time-delay and the enormous amount of energy it sucks out of them goes straight into that baby. Here we have, perhaps, an intriguing possible answer to the question of 'why monogamy?': a demanding baby. A helpless infant, one that takes forever to grow, is at greater risk for longer from homicidal males, whether they know what they are doing or not. Perhaps a demanding baby may also keep males occupied when they would have otherwise been on homicidal sprees, which is why we see almost no infanticide in pair-bonded primates today and must propose it as an ancestral state through computational models. Whether the highly demanding, totally dependent baby came before the pair bond or not remains firmly in the semantic realm of poultry-based idiom. What is certain?

Our babies are costly, weird and absolutely useless – and it's a minor miracle we have them at all.

---

* Or finishing the edits on the manuscript for this book.

## CHAPTER FIVE

# Georgie Porgie, Pudding and Pie: Conception and Fertility and Fat

Georgie Porgie, pudding and pie,
Kissed the girls and made them cry,
When the girls came out to play,
Georgie Porgie ran away.

Despite what your high school sex education teacher might have told you, in humans, successful pregnancy after sex – even timed between perfectly fertile partners – only works out about 30 per cent of the time. And that's only while you're young – female fertility takes a dramatic swan dive right off the plateau of normal primate reproduction. Men have it either easier or harder, depending on how you feel about having small children while aged.[*] While quite a lot of men can still produce viable sperm up until they are no longer viable themselves, for some men age does reduce the quality parameters of their offering. Primate 'fecundability', which is a slightly annoying way of saying their ability to have a baby, is based on several factors, however. You need the eggs and the sperm, of course, and the mating system to bring them together, all of which we saw in the previous chapters. However, there are several factors in conception, actually bringing it all

---

[*] Let me help you – no one feels good about this. They feel tired – so tired. The only saving grace is that they would also feel tired if they were not aged, because small children are essentially energy vampires.

together in baby form, that mark out humans as unique: when we breed, how often and what it takes to get us there.

It is, in fact, a minor miracle that humans even manage to reproduce at all when you consider that we are actually terrible at getting pregnant – about 50 per cent of menstrual cycles are a bust, and when we do it right, our chances are only about 30 per cent. For most other primates, healthy individuals mating during the right time results in a pregnancy a whopping 90–95 per cent of the time. Like humans, most primates have a roughly 30-day menstrual cycle. However, not all species have them all year round, so their success rates come on top of a very limited window of opportunity.

Seasonal babies are a sensible idea, evolutionarily speaking, if you're reliant on one bumper crop to see you through. The lemur species of Madagascar are seasonal breeders, for instance, because Madagascar has very intense seasons where food is either abundant or not. Seasonal breeding is common in animals and many species of primate bunch up reproduction all at the same time. That way babies are born to catch the wave of most available food either when mama needs it most, at the end of pregnancy and early breastfeeding, or when baby needs it most, when they move off the breast and on to whatever else is around to eat. Seasonality can really put a pinch on the time available to reproduce, but so can other pressures, like predators. Adorable and highly edible squirrel monkey babies in any given troop are almost all born within a week of each other, around February–March. This is uncoincidentally also when the largest number of hungry birds are seen circling overhead, looking for delicious baby squirrel monkey snacks.

'Oestrus synchronisation', or having fertility cycles sync up, is the mechanism for ensuring you only have one week of birthday parties every year. You may have heard that human women do this too, for evolutionary reasons that are always just slightly undefinable. There's only one problem with this seemingly evolutionarily backed idea: it's not actually true. Modern life has intervened to give us a fascinating new data set: the timing of menstrual cycles recorded in a 'period tracker' app called Clue. Despite the seemingly unkillable myth that women who spend time with each other synchronise their reproductive cycles, data from a few thousand women on the app supported what many smaller studies had already suggested. Women do not synchronise their menstrual cycles, perhaps because predatory birds aren't circling overhead and trying to eat us every February. Women might, however, synchronise the human adaptation for confirmation bias and unreliable recall that might make them think they do.

Humans do, however, have a breeding season. If you grew up looking around your primary school class and quietly resenting all the kids that you had to share a birthday with, there is actually more to it than just a cupcake-depleting coincidence. In other animals like our friends the lemurs and the tasty squirrel monkeys, environmental pressures like seasonal differences in food abundance and the presence of predators encourage seasonal breeders. And how do you know it's the right season? The same way you know the season has changed, by looking at cues like levels of sunlight. In climates where levels of sunlight vary throughout the year, birth rates show strong peaks and troughs. Birth seasonality is much stronger in the northern latitudes – where lack of sunlight is a thing – than the cosy equatorial regions. Comparing the birthdays of babies way, way up at 70.7°N on the island of Ulukhaktok (in Arctic Canada) with those living in the rather more temperate Papua New Guinea at 4°S shows a remarkable difference in birth season. But it's not all about the weather outside – or if it is, then it's about what humans feel like doing in that weather. The Ulukhaktok babies appear at the tail end of winter into early spring, meaning they have spring/summer conceptions that reveal a lot more about what the parents are up to in nice weather than some evolutionary dictate. There is no 'one' time to conceive a human baby. The birthdays in the study of Papua New Guinea had nothing like the Ulukhaktok pattern, but they revealed just as much about the society that was doing the conceiving as the Arctic example. One yam-growing and, quite frankly, yam-fetishising community had such strong proscriptions against sex during the months of the critical yam harvest period that even *joking* about it was banned. Unsurprisingly, they had a strong peak in births in October – nine months after the sex ban ended in January.

In the urban societies of the northern hemisphere, births are pretty seasonal, mostly occurring from July to September. October is the key time for baby-making in the northern latitudes, resulting in July babies. This seasonality slips as you travel southwards: September to December births are more common in the tropics, and January to early summer births near the equator and in the southern hemisphere. But we humans are nothing if not ruled by our cultures, so there is a shocking number of Leos and Virgos around the world because of the timing of the Christmas/Midwinter/New Year holidays, no matter where they're born (see Figure 5.1). Season of birth even fluctuates with class and education in our big urban societies – there's evidence that parents who can afford to be choosy time their offspring for an optimal start to formal schooling, for instance. How exactly these clever

| Latitude | Jan | Feb | Mar | Apr | May | Jun | Jul | Aug | Sep | Oct | Nov | Dec |
|---|---|---|---|---|---|---|---|---|---|---|---|---|
| 60–70 | 8.3 | 7.1 | 8.3 | 8.2 | 8.2 | 8.4 | 9.1 | 9.0 | 8.9 | 8.5 | 8.1 | 7.9 |
| 50–60 | 8.6 | 7.7 | 8.2 | 8.2 | 8.4 | 8.4 | 9.3 | 9.1 | 9.0 | 8.0 | 8.1 | 8.2 |
| 40–50 | 8.2 | 7.4 | 7.7 | 7.6 | 8.3 | 8.5 | 9.2 | 9.3 | 9.4 | 9.0 | 8.1 | 8.4 |
| 30–40 | 9.1 | 9.5 | 9.0 | 10.4 | 9.3 | 10.3 | 10.3 | 11.1 | 11.7 | 12.0 | 10.9 | 10.7 |
| 20–30 | 8.6 | 7.3 | 7.9 | 7.4 | 7.8 | 7.6 | 8.4 | 9.0 | 9.0 | 9.3 | 8.8 | 8.9 |
| 10–20 | 8.5 | 7.3 | 8.0 | 7.9 | 8.3 | 8.0 | 8.3 | 8.7 | 9.0 | 9.1 | 8.5 | 8.5 |
| 0–10 | 8.5 | 7.1 | 8.4 | 8.2 | 8.5 | 8.1 | 8.4 | 8.5 | 8.7 | 8.8 | 8.4 | 8.3 |
| -20–-10 | 7.8 | 7.6 | 9.1 | 8.6 | 9.2 | 8.8 | 8.6 | 8.4 | 8.2 | 8.0 | 7.8 | 7.8 |
| -30–-20 | 8.6 | 7.8 | 8.8 | 8.3 | 8.5 | 8.3 | 8.3 | 8.5 | 8.7 | 8.0 | 7.7 | 8.3 |
| -40–-30 | 8.6 | 8.0 | 9.2 | 8.4 | 8.7 | 8.3 | 8.2 | 8.3 | 8.1 | 8.1 | 7.9 | 8.0 |
| -50–-40 | 9.3 | 7.6 | 8.1 | 7.5 | 10.3 | 7.2 | 9.1 | 8.4 | 7.6 | 8.5 | 8.7 | 7.5 |
| -60–-50* | 15.4 | 11.5 | 7.7 | 0.0 | 7.7 | 7.7 | 11.5 | 3.8 | 15.4 | 11.5 | 0.0 | 7.7 |
| Grand total | 8.6 | 8.2 | 8.5 | 8.9 | 8.8 | 9.0 | 9.3 | 9.6 | 9.9 | 9.9 | 9.1 | 9.1 |

Figure 5.1. *Seasonality of birth across different latitudes. Data by country, based on location of capital, from UNdata, United Nations Statistical Division (retrieved 2022; the asterisk indicates a small dataset). Note there are no entries for 0-10°S.*

parents manage that is another question altogether, however, because the other thing about human conception is that we are *terrible* at it.

Even if we do manage to get pregnant, we are still much less likely to have a baby than our primate relatives. About 20–30 per cent of our sperm and egg mergers last less than five weeks, a short enough time that many women may not ever know that they were what is termed 'chemically' pregnant in the first place. A further 30 per cent of pregnancies will be lost to miscarriage. Baboons, by contrast, who cannot even operate a pregnancy test, will have about 85 per cent of those early implantation events result in a bouncing baby baboon. Humans actually have a remarkably high rate of failed pregnancies, or miscarriages, and are one of the only species[*] to experience specific pregnancy-related health conditions like pre-eclampsia, where your blood pressure goes out of control in pregnancy, and you can end up seizing and

---

[*] Us, one type of monkey and guinea pigs. Go figure.

rather bathetically dead if you are in any sort of period drama set before the 1920s.* So, why we are so bad at babies? The answer is twofold, and one more or less solves the other: we are bad at getting pregnant in any one menstrual cycle, but we have a lot of menstrual cycles to try in. Having periods all year round is a good way to balance out that rubbish success rate.

Clever of us indeed to have so many – it suggests on one hand that humans are fairly resilient to the stresses of our environment, compared with our primate relatives. Chimps walk a thinner line than humans in terms of food availability; bad months are bad to the point where it's simply not possible to have a baby because the mother's body has gone into 'amenorrhoea' – she has shut down reproductive function and is not releasing eggs. Human women well-nourished and in good health today, however, can expect to have around 450 menstrual cycles – periods – over their lifetime. This is actually a massive increase on what we might expect if we were toeing the reproductive line a little more aggressively. If we were to live in one of many societies that keeps a non-supermarket-based way of life, we might breastfeed for several years rather than several months, have a lot more than our current average of one or two babies, and spend a lot less of our lives having periods and a lot more pregnant and/or breastfeeding.† The Dogon, non-supermarket-based farmers who live in Mali have what's known in demographic terms as a 'natural fertility' regime, which just

---

* RIP Lady Sybil of *Downton Abbey.*
† Having spent the entirety of writing this manuscript doing one or the other, I can assure you is not an improvement.

means they avoid contraception you have to pay for.* They have more like a hundred periods over their entire reproductive lifetimes. Keeping ourselves awash in the ovarian hormones that spike every time we have a period also increases the risk of diseases that are sensitive to hormonal factors, like breast cancer. As early as the seventeenth century physicians were making the connection between the (allegedly celibate and childless) women who became nuns and their increased risk of breast cancer.

Whether we are experiencing them or not, the fact that we are capable of having this monthly menstruation cycle is still pretty odd. No other animals get periods the way modern human women do; menstruation isn't even a thing in most animals. We, and our Old World cousins – the monkeys without tails and the great apes – and a shrew and two types of bat are the only animals in the world that see fit to regulate our reproduction by bleeding off component parts. Why we should do so has been subject to considerable speculation. First, of course, you have to get over the fact women have periods at all, which, according to the highly engaging *PERIOD* podcast hosted by the also highly engaging anthropologist Kate Clancy, may or may not have happened yet. You also have to overcome some serious cultural issues – and, no, not the taboos like the infamous 'menstrual huts' that early ethnologists spent so long taking notes on, but right here and now in developed countries.

The idea that women are unclean during certain phases of their menstrual cycle has been 'scienced' quite a bit

---

* Natural fertility is a terrible term for humans because humans mess about immensely with their fertility, whether through sexual practices, exciting chemicals, cultural taboos or whatever other tools they've got to hand. Messing with our nature *is* our nature.

over the years, with rather interesting assertions that menstruating women exude a 'menotoxin' – some noxious, venomous substance that fouls crops and gives babies asthma and colic.* This was all kicked off by a rather novel experimental study of making women on their period touch flowers, and then women *not* on their period touch flowers, and seeing what happened to the flowers.† This is not necessarily the latest word in scientific practice, though still a step up from the (male) researchers who decided to uncover the evolutionary role of menstruation by working out how much lap dancers got tipped throughout the month.‡ However laughable it may sound, this toxic idea has been extended to understanding the evolution of human reproduction as well. Menstruation that served to kill off lower-quality sperm, protected against other conditions as part of a cycle of immunity or 'chose' viable embryos, would indeed be something slightly more evolutionarily useful than the ability to do low-grade damage to a dining table centrepiece once a month.

It may come as a surprise to certain scientists of the twentieth century,§ but women do not actually secrete a toxin while menstruating and we are no more toxic to sperm or accompanying pathogens than to flowers. The idea of a 'choosy uterus' is quite an interesting one, however. It suggests that women have a mechanism inside

---

* So, basically, anything you could have been burned at the stake for.
† As Clancy points out, maybe the nurse who originally came up with the 'I can't touch flowers right now' excuse just didn't want to take time away from her actual job to faff with some doctor's thank-you bouquet. And for this, we get 80 years of misdirected science.
‡ And this is why we need research ethics committees.
§ Or Witchfinders of the seventeenth century.

them for determining how much energy we are going to invest in any given pregnancy. It also suggests that there is a reason for being choosy, because not all of our sperm + egg combinations are high quality. This may be what lies behind our terrible pregnancy rates: we just don't produce as many viable embryos as other species. We – like old-world monkeys, elephant shrews and those bats – have sex whenever we feel like it, as you recall. This means by the time we get around to fertilising our eggs, they might have already ripened and sat around for a bit, and be a bit more prone to genetic errors. If we let non-viable embryos bathe in our blood – which is what human, primate, tree shrew and those bat embryos do – then, frankly, it's a waste of blood. Better to slough off the whole lot and start again. It's also possible that the whole period thing is just a pre-emptive strike by the mother in the war for her uterus, given the whole blood-bathing invasive nature of our embryos. In animals that don't get periods, just the presence of the embryo kicks off the preparations for pregnancy. In animals like us (and the monkeys, shrews and bats), the mother has a quality-control mechanism that empties out the potential pregnancy-supporting tissue if things aren't looking good. So it seems that all the monthly bleeding and inconvenience is just an annoying side-effect of our demanding quality standards.*

There's one final major hurdle† to getting pregnant and that is you need an adult female. Remember our big male, small female conundrum? Well, adult females,

---

* Congratulations – if you're reading this, you were a high-quality embryo.
† Biological hurdle. The cultural ones are between you and your online-dating profile.

while seemingly littered all over the place, do not get to be adult females without considerable investment in the first place. This is how we come to the second critical aspect of life history: choosing when to switch from growing ourselves to growing the next generation. Not only can life history be fast or slow, but the point at which you make reproduction possible is a moveable feast and this is what determines most of how animals on earth live – including us.

We've seen in Chapter 2 that there are some (sketchy) rules for how fast and slow animals live; unsurprisingly, similar theorising has given us the idea of two life-history strategies in terms of childhood. On the one end, you have your short-growth, long-reproducing animals. On the other, the long-growth, short(er)-reproducing ones. For instance: cats are keen reproducers, if you let them, and many a fur-baby has surprised its owners by turning up pregnant at the tender age of six months. A cat is ready to reproduce after about 3 per cent of its time on earth.* Greenland sharks, however, wait until a more respectable 150 years or so, passing a good 37 per cent of their admittedly leisurely 400-year life span in a pleasantly childfree bubble.

And what about us, and our primate relatives? We mature around 15 and live to maybe 80, so that's a good 18–20 per cent of our lives we waste being children and not making more humans. Orangutans, not renowned for their dynamism, get to reproductive age sometime between 6 and 11. Gorillas, chimpanzees and bonobos all kick off

---

* That's for a house cat who lives around 15 years, not a feral cat, which lives about half that time. My cats Enkidu and Gilgamesh, however, are going to live to be at least 30, because I will not hear of anything else.

around 7 years – but only in captivity. In the wild, chimpanzees and bonobos start their reproductive lives *years* later, closer to 9 or 10.

And this is where it starts to get interesting. The length of our childhood – how fast we can get to reproductive status – isn't set in stone. It's changeable, adaptable. And if it's adaptable, you can bet your bottom dollar that we have adapted it – and our whole life-history strategy along with it. That's what we're talking about in this book – what humans have done with the cards we've been dealt, and how we're still tinkering with our basic operating parameters. So, what are the levers we are pushing and pulling to change this basic aspect of our life-history strategy?

Well, first off, in order for a primate with female reproductive organs to reproduce after mating, she has to be two things: sexually mature and capable of maintaining a pregnancy. Both of those things require energy and are a lot more dependent on the environment than you might think. Sexual maturity in primate females, for instance, is not just a function of age or genetics. It is very closely tied to levels of body fat and nutritional status.

Young female rhesus macaques* fed a high-fat diet reached sexual maturity months earlier than their normal-diet counterparts. They didn't get fatter, they didn't grow bigger, but they did achieve reproductive status earlier – because

---

* The astute reader will note that it is almost always rhesus macaques in laboratory research. This is because they are the most common monkey model used for a host of human behaviours, both social and biological, due to an unfortunate (for them) balance of being biologically similar enough to get our diseases but very cheap to, err, replace. So, when you read 'lab monkey' it is likely to be exactly the type of little brown monkey you are picturing in your head right now.

evolution would rather have you making babies than bulking out. The opposite also applies: two orangutans with eating disorders had the onset of their first menstrual cycle much later than an obese one. Animals in captivity generally reach reproductive status much earlier than their wild cousins, a feat that is usually attributed to the lazy lives led by luxury captive apes compared with the lazy lives with bursts of occasional terrifying violence, monkey-hunting or social sex led by apes in the wild. Interestingly, this is not true of gorillas, who are ready to make more gorillas at the age of seven, no matter whether they live in zoos or in the wild. This suggests that, despite the constant threat of predation and their endangered status, gorillas in the wild are actually hitting peak gorilla energy balance. Of course, it's hard to overdo it when all you do is leaves.

A certain amount of body fat seems to be necessary to trigger the hormonal changes that kick-start puberty, the changes associated with turning a kid into a potential reproducer, in primates, or at least in females. This has been linked to the fact that fat tissue is one of the few places you can get the hormone oestrogen from besides the reproductive organs themselves – and greater reserves of fat might mean the potential for higher levels of oestrogen. Girls with higher body fat generally have an earlier first period, something that we like to think makes evolutionary sense in the same way seasonal breeding does – you don't want to take on baby-making without the energy budget to do it successfully.

Generally, there is a widespread idea that the way humans live today – deep in the black in terms of our energy budget – is behind the falling age at which girls get their first period. We suspect that in the past, girls would have hit puberty much later than they do today, because when we look at groups that are living without a ridiculous surfeit of energy in Fruit Loop form, girls tend to have their first period on average around the age of 12. In big studies run in developed, Fruit-Loop-friendly economies, there seems to be a link between childhood obesity and the onset of puberty in girls, something that was developed by the rather impressive scientist Rose Frisch[*] into the 'critical body weight' hypothesis. She thought there was a lower limit for body fat and once you crossed that threshold the body would signal the reproductive system to turn itself on. Her studies of female athletes led to the widely accepted idea that a body fat percentage of 17–22 per cent is necessary for reproduction in human females and that the earlier this is achieved the earlier the onset of reproductive ability.

More recent genetic research has identified a variety of locations in the human genome that are associated with when a girl gets her first period; many are thought to regulate functions related right back to hormones and body weight, so perhaps there is a bit more feedback in the system than the easy model of calories in, hormones out. Male reproductive maturity, comprised of slower-moving changes to gonad size and hormone production, is more difficult to identify as an event than in females. It's also much less studied, for reasons that I suspect are deeply embedded in the cultural hang-ups of the kind of societies that carry out reproductive research. For whatever reason, it has been much more difficult to identify the actual factors that trigger the launch of reproductive hormones and a million wispy moustaches, so the jury is still out on the main factor in the timing of

---

[*] Born in 1918 to a family of Russian immigrants in the USA, Frisch (then Epstein) went on to marry, have children, work on the Manhattan Project, and revolutionise the study of women's fertility and its connection to body fat in her long career at Harvard. As you do.

male puberty.* However, studies of large numbers of boys over time have found that childhood weight is linked to puberty, just as it is in girls: the more of it, the earlier it happens. Boys lag behind girls by a year or two, but tracing ages they hit their pubertal stride shows us that as people trend towards having more fat on them as children, puberty comes earlier and earlier.

It may be that what you eat is almost as important, however, as how much of it you are shovelling in. In China a large survey of kids found that a higher-quality diet was associated with a later puberty – but that body fat was still having some sort of effect. Building off big studies elsewhere in the world, this supports the idea that what you eat – rather than just a straight-up 'how much' – determines how old you are when you become capable of reproduction. A diet with a little less meat and cola but more fibre and isoflavones – the magical oestrogen-like stuff produced by legumes and nuts – seems to set a young person up for a longer period of growing their skeleton and other bits before swirling down into the energy sink of reproduction. A study carried out by Taipei Medical University researchers found that for every extra gram of animal protein eaten a day, the age of a girl's first period moved forward two months. This is something of a cycle, it seems, as mothers who hit puberty at an earlier age go on to have bigger children – and they are more likely themselves to have an earlier puberty. Somewhere in us is an insistent little voice telling us to accrue fat – so we can have more babies.

---

* Presumably having approached male puberty by knocking repeatedly on their door and then deciding, after careful consideration of the odour and noises coming from within, that discretion is the better part of scientific valour.

This obsession with fat doesn't just end in childhood.* Females don't just need fat to reach sexual maturity. They need it to stay capable of reproducing – of carrying a pregnancy successfully. Women with eating disorders, breastfeeding women and professional athletes with low body fat percentages – particularly gymnasts and ballet-dancers – frequently suffer from a type of calorie-deficiency where periods stop completely. Below a certain level of body fat – that 17–22 per cent mark – women cease to ovulate and the menstrual cycle shuts down. This can be successfully reversed in most cases by decreasing energy expenditure and gaining body fat, but the fact that the whole system essentially powers down under duress points to the importance of body fat in human reproduction.

Not only do we have critical levels of body fat for reproduction, we need much, much more fat than other primates. Our female rhesus macaques from the lab hover between 8–18 per cent body fat; human females struggle to reproduce with less than 17 per cent. That's nothing, however, compared with chimpanzees – a recent study found female chimps had a body fat percentage between 0–9 per cent, and they're far closer to us in terms of evolutionary history than macaques. By sneaking around Amboseli National Park in Kenya with a blowpipe full of tranquilisers, researchers discovered that female baboons manage on about 2 per cent in the wild. However – and here's a caveat you want to keep hold of when you think about adaptation – female baboons that have managed to secure a cushy living from a human garbage dump behind the Lodge Hotel have a hefty 23 per cent body fat. And

---

* If it did, what would explain bar snacks?

guess what they spend most of their time doing with all that extra energy? Having more babies.

This points us to the next thing we need to understand when we talk about making humans: why on earth do we need so much fat? One argument for our having generally more fat on and about our persons than other primates, and not just during reproduction, is that humans are used to living with uncertainty. Our food sources throughout our evolution, wide and varied as they are, do not occur with seasonal regularity, and our main ecological niche is essentially finding new things to eat and new ways to eat them before we die. It is a far safer idea to carry the energy you need around with you than to assume your environment is going to cough up. We can see a little of how this works in the seasonality, or lack thereof, in primate reproduction, when the canny squirrel monkey times its babies to coincide with the bonza food season because it isn't carrying around the fat we are.

Storing body fat has its advantages. Polar bears, for instance, have an extreme feast-or-famine existence. The immense amount of fat female bears are able to store up during the feast months is what allows them not only to survive the lean ones, but to meet the energetic costs of reproduction during hibernation.* For the huge numbers of species whose seasonal menu ranges from 'lots' to 'uh-oh', storing body fat is the way to maintain energetic balance. However, there are downsides. Big bodies cost a lot of energy to move around, for one thing. And, as noted very politely by researchers observing some rather pudgy macaques in trees, carrying extra can have unfortunate consequences for what they call 'terminal branch feeding',

---

* There is something to be said for an adaptation that allows you to go through pregnancy asleep.

or nibbling down at the thin end of the branch that held your weight in the dry season.[*] Those same researchers point out that while fat might be fine for lab monkeys, wild macaques rely on their lithe and limber status to exploit all the sources of food they need to stay alive. They simply can't afford to invest in fat, whereas we have thrown ourselves into it wholeheartedly. And the number one beneficiary of our fat stores? The next generation.

---

[*] This book contains an unforeseen but unavoidable number of monkeys falling out of trees, for which I apologise.

## CHAPTER SIX

# Bake Me a Cake as Fast as You Can: the Joys of Gestation

Pat-a-cake, pat-a-cake, baker's man,
Bake me a cake as fast as you can ...

We discussed in the previous chapters the variety of schemes primates have developed in order to generate offspring. There was quite a lot about male competition, infanticide and the business of pairing off, and we generously deigned to allow females a look in towards the end. It was quite the struggle to get through conception, and we hinted at some of the levers that might be pushing male and female animals towards different lengths of time spent growing. However, there was not a lot about the actual business of growing babies, which, as you might imagine, does have some bearing on how childhood works. The timeline for life-history trade-offs that affect our childhoods extends from even before the point of conception and, as later chapters will discuss, into a sort of infinite cycle of dependency in some of our immediate relations.* But of all the choices in where to invest in our species, some of the most clear-cut life-history variables are the time spent draining the old generation in order to build the next.

Pregnancy – how we do it, and for how long – is adaptive. Remember the earlier discussion in Chapter 2 about altricial – helpless – baby rats versus precocial, ready-to-go

---

* Your parents know exactly what I mean.

young giraffes? You can invest early, during pregnancy, growing that fully cooked giraffe for ages and ages, or you can delay the investment for after birth, like the little rat pups that are going to need all the energy a pizza can offer. Parental investment is not an either-or scenario though. As you may have guessed from your own experiences and/or casual acquaintance with our species' young, we are in the altricial, useless baby category. But why? What on earth could have possessed us, trundling around the evolutionary landscape, to favour big, heavy, hairless babies? This is not the sort of thing you imagine comes in handy in a cave lion fight.* This chapter will look specifically at the investment required to take us from conception to actually giving birth to a child that, one hopes, will grow up human – and why we shut it down at such an early stage.

This chapter was very nearly going to be a much more introspective look at pregnancy, because it was written by a pregnant author, and my assumption was that I would be utterly fascinated by the subject of pregnancy while experiencing it and wish to write about it. However, only the fact that pregnancy is evolutionarily extremely important came even close to compelling me to revisit the subject on the page.† Other women may find being pregnant a fascinating, emotional, special time of life, but I found the entire experience fell within expected parameters‡ and

---

* It is my increasing concern that the way we think of human evolution has become a lot more like how we think of mixed martial arts and a lot less like the actual business of living. Despite our tendency to view evolution as a contest, it's not like there was some announcer shouting 'FINISH HIM' in a last epic cage fight with the Neanderthals.

† That, and the threat of having to repay my book advance.

‡ I expected it to be tedious and uncomfortable and, lo, I was not disappointed.

was such a long, drawn-out process that it could barely keep my attention as a narrative, let alone anyone else's. As I tried to find anything of interest to convey the critical, highly adaptive nature of human gestation, I ended up mostly thinking about how much I missed American deli foods and, as time progressed, my feet.

And that, it turns out, is most of the story. Human pregnancy is not much of an event, if you are lucky. It is a nine-month marathon of annoyances that, in my case, were thankfully small and didn't have much of an impact on anything other than my waistline.[*] This is not always the case, however. Human pregnancies do have a few exceptional characteristics, one being that they have the not-terribly-adaptive-seeming potential to kill you. Also, they are made possible by what must be one of the weirdest adaptations in animal life ever made.

We are placental mammals, which is a way of defining ourselves against egg-laying oddities like the duck-billed platypus and more common marsupial mammals like kangaroos, Tasmanian devils, and the deeply misunderstood and scientifically mispronounced North American (o)possum. A placental mammal is a very weird thing to be, even odder than an egg-laying mammal. Eggs have been around forever, after all. Adding milk to eggs, which is how the platypus gets its status as a mammal, is even technically kosher.[†] The placenta, by contrast, is not present

---

[*] Shout out to Abraço café bar for serving me the Turkish pastirma simit version of a pastrami bagel. A lot.

[†] It is not, however, kosher to eat a platypus, because if you think of them as water-dwelling then they don't have fins and scales, or if you think of them as land animals they go on their bellies and don't have split hooves or chew cud. I told you I spent a lot of time thinking about food.

right down the animal-evolution line. It is a little organ
that forms in placental mammals when they get pregnant,
and only when they get pregnant, to act as a sort of filter
between the mother and the embryo that otherwise would
be identified as a foreign body and summarily expelled.
Placentas are necessary for the prolonged siege that an
embryo makes on its mammal mother, which is why
marsupials like possums have tiny little babies in big protective pouches – their rudimentary placentas simply can't
support a foetus for more than a few days, so their babies
have to be born into a sort of uterine limbo in the outside
womb of their pouches. Placentas filter nutrition, waste
and things that might damage the foetus; they also import
the maternal antibodies that protect the baby and keep it
bathed in a constant stream of blood-borne nutrients,
Bathory-style.*

   So far, so weird – but wait. The placenta isn't even,
strictly speaking, human. The genetic code that tells your
body to make syncytin, the substance that creates the
placenta, is actually from an RNA virus. Some time around
130 million years ago, something in us captured the bit of
something we now call the human endogenous defective
retrovirus (HERV-W) and harnessed its ability to make
syncytin into the ability to make placentas, so we could
balance growing our young inside us, instead of outside in
tasty egg-shaped predator snacks. Co-opting viral DNA
definitely strikes one as odd, though it is not actually so
uncommon – if viruses can adapt us to suit their needs, it
seems only fair we occasionally return the favour. What's
more is apes like us have very odd placentas indeed. Not

---

* While the historical fact of Countess Elizabeth Bathory of the
Kingdom of Hungary as a sixteenth-century vampire bathing in the
blood of virgins has been contested, the image does linger.

only are they creepy adapted RNA viruses, they are also sending all sorts of coded commands to the mother's body. Placentas in primates – and only in primates – are responsible for hormonal signalling telling the mother not to expel the foetus. Primate placentas regulate something called corticotropin-releasing hormone (CRH), which is not anything you need to remember except that it affects when a baby is born. The timing matters here: if you make all of your CRH in the placenta, instead of up in the brain like a Pizza Rat, babies are born later.

This, it seems, is part of a pattern that comes into sharp relief in human pregnancies. In us, the placenta isn't just some filtration system, it's an *infiltration* system. The placenta attaches right up into our uterine wall, plumbing directly in to give baby far greater access to the nutrition it needs to grow – and far greater control over the mother than any other primate species. Human placentas regulate the flow of nutrition in and out in a remarkably non-even-handed way, favouring the foetus over the mother in ways that can, if they go wrong, kill us. In the very beginning of a pregnancy, the embryo starts to invade and remodel the blood-flow system, creating a little blood-drenched haven for itself that the mother's body can no longer restrict or control. Placental control of the main arteries that send blood to the baby mean that, if the baby ever feels it is not getting the nutrition it needs, it can start demanding more by pulling more blood through. Mama's blood pressure goes up and, if not checked, she can go from hypertension headaches to the swelling of pre-eclampsia straight through to seizure, stroke and death.

The human placenta is very fond of sending the mother hormonal signals about what the baby wants. It creates human placental lactogen (another long name you don't need to remember) to tell the mother's body to produce more tasty sugars after eating, the better to grow with; if her body tries to clamp down on this too hard by producing more insulin, she ends up in a sugary war of attrition called gestational diabetes. Constant thirst, constant urination, exhaustion – sure it sounds like normal pregnancy, but that imbalance can lead to risk of death or serious problems ranging from a baby that is too big – including the aforementioned pre-eclampsia. Gestational diabetes is a pathology literally caused by your own baby and can be debilitating for the mother as well as potentially fatal for the baby, but somewhere along the evolutionary line we decided that extra pressure on growth in the womb was worth it.

This, on the surface of things, sounds like a terrible, invasive, drawn-out way to go about having a baby. It is also, below the surface of things, a terrible way to go about having any old random baby. What it is, however, is the only possible way to have a human baby, one that fits our particular brand of big and fat and helpless. This brings us to the second thing that is remarkable about human pregnancy, the thing that I found mattered to me increasingly more while unable to drink through the onset of the very long drawn-out global pandemic in 2020: the nine-month gestation period of a human animal is a lie. Human pregnancies are not nine months. They are nine-months-that-are-actually-ten-months, forty weeks on average, with two weeks either side being completely normal. Even then, we have done that thing again where, if you plot us on a chart of animal life history, we come out – you guessed it – weird.

All mammals evolved to have a half-in, half-out baby-preparation process, with part of the investment in the baby being made while it is in utero and part after they are born – this is the point of milk. The length of mammal pregnancies is designed to accommodate this. One of the ways we can characterise parental investment is by charting the time we take letting the baby co-opt the mother's physical being – the duration of pregnancy. Thinking back to the fast-versus-slow life histories in Chapter 2, we can generally see big, slow mammals having big, slow gestations, and small ones having shorter waits. One Ton Giraffe Mum is going to take around 430 days, or a year and two months, to create her masterpiece, while our good friend Pizza Rat churns out her 5–12 pups – also called kittens* – in 20 odd days. All the while, there we are, trucking along with our nine months,† waiting on a baby that has so little physical ability when it is born it can't even sort moving its head out.

The broader pattern in mammals is that the larger you are hoping to be when you grow up, the more 'finished' you are when you're born; baby animals hoping to be very large adult animals are born more ready for the world than those not eyeing such great heights. Little animals like our rat friend are putting a lot of effort into pregnancy to grow their babies quickly, while the bigger animals don't have to devote quite such a proportion of their resources to their pregnancies. That in turn reminds us of the other thing we learned about baby rats in Chapter 2 – their mother cannot quite manage to store all of the energy her babies need to reach functional ratness. Pregnancy is an investment – a dangerous, resource-draining investment – in the next generation. Big animals are playing the long game, doubling

---

* Yes, baby rats are called kittens, in an irony that is lost on everyone except for, now, you.
† Forty-two might be the answer to life, the universe, and everything, but I tell you from experience it is two weeks too many for a pregnancy.

down on a pass-line bet that their expensive pregnancies will result in a viable baby, while small ones have scattered their bets all over the craps table in the hope that at least some of them are going to come good. This means the giraffe has gone all in on her year-plus, while Pizza Rat is only betting a little less than a month.

So, the classic answer to why we have such long pregnancies is that primates are difficult to grow: we use that extra time to get them the energy they need. That long gestation gives us time to grow that foetus slowly. Remember how in the last chapter we talked about the critical nature of human fat storage in being able to have babies? Well, in pregnancy it's all about how efficient you are at getting those fat stores to the demanding little interloper in the womb. You will have heard of 'eating for two' – and also probably by now that 'eating for two' is more an aspiration towards joy than a strict biological necessity.* Lots of people around the world actually decrease their calorie intake, particularly in the third trimester when things get a bit crowded. The extra caloric demand of pregnancy is shockingly actually more of a theory than a proven fact, because it is very difficult to work out the constantly fluctuating mass and activity levels of a pregnant woman as well as keeping exact track of what she is eating. The official best guess from developed Western countries averages out at only about 200 calories per day in the third trimester – that's less than one scoop of ice cream.†

---

* I really cannot emphasise enough how much I, millennial Californian (failed) vegetarian, enjoyed the cheeseburgers and pancakes with sausage and bacon that I ordered from the café on a near-daily basis.
† ... but *ten* pickles.

We all have ideas about what that means for pregnancy weight, of course, some more realistic than others,[*] but the point of our fatty females is that they can carry extra resources in convenient locations around hips, bust and thighs for maintaining a pregnancy even when the environment stops providing peanut-butter ice cream. Unlike Pizza Rat, we can carry enough energy on our persons that we are less vulnerable to the kind of nutritional crisis that leads other animals to spontaneously abort or even resorb developing foetuses. There are thankfully very few instances of real-world testing of calorie restriction in pregnancy. The best known is the effect of the Dutch Hunger Winter of 1944–45, where the blockade of the Netherlands led to urban starvation and pregnant women were thought to be getting somewhere between 1,400 and 700 calories a day as conditions worsened. Women getting fewer than 1,500 calories a day (out of a recommended 2,000) in their third trimester, either losing weight or gaining less than a half kilogram a week, had babies that were underweight by around 300g, but they still had babies. We appear to be built to stay pregnant even in the face of limited or fluctuating resources, and if that means erring on the side of ice cream, then so be it.

Metabolic conditions determine energy storage and the potential for energy transfer – and those conditions aren't entirely set in genetic stone, but are determined in part by

---

[*] The much-touted ability of female celebrities to regain a 'pre-baby body' in the weeks or few months after birth for some reason usually fails to mention the availability of paid staff to take care of the nutritional, emotional and physical needs of a new mother. It is my advice that if you wish to look like you have not just had a baby when you have in fact just had a baby, you should probably be very wealthy. I find this is good advice for most things in life.

environment and diet. We are what we eat, after all, and what we eat also seems to have an effect on how our babies grow in the womb. Primates, in their globe-spanning determination to adapt and thrive, cover all of the possible niches and all of the possible diets. There are near-exclusive insectivores, like the big-eyed, small-brained lorises[*] down at the base of the primate family tree, and highly specialised folivores – exclusive leaf-eaters – just like the gorillas[†] way at the other end. For quite a long time there was a clear line drawn from the protein-heavy primates to the ponderous chewers to the fruit pickers dependent on seasonal bounty – the guys with the good protein sources grew fastest, while those who had to work harder on acquiring calories necessarily drew out the time they took growing their babies. This goes some way towards explaining the primate reversal of the normal mammal pattern of 'small gestates fast, big gestates slow' in primates – it is the little guys in our order that eat the most protein-heavy diets, while the larger animals are far more likely to opt for leaves or fruit. They are the ones pumping in the calories, so they should be the ones building their babies fastest.

The primacy of straight calories in deciding how long your baby takes has been challenged by studies showing that leaf-eaters in both the prosimian lemurs and the anthropoid colobine monkey gestate proportionately longer than sugary, calorific fruit-eating lemurs or colobine-sized fruit-eating macaques. This is unexpected – unless what counts is quality of calories and what they come from. Leaves are a better source of protein than fruit and, what's more, they are available year-round,

---

[*] This is not why they are called slow lorises, by the way. They just move very, very slowly.
[†] Gorillas *and* adult human women in stock photography.

unlike seasonal fruit. When it comes to quality protein, tarsiers aren't the only primate to instil terror in the other smaller, snackable animals in their environs. We know that chimpanzees hunt and other species will scavenge animal protein, but neither chimps nor that apex predator, humans, get anything like the percentage of their calories in protein as a tarsier (even modern Americans, evangelists of the animal-protein-consuming world, only manage about 36 per cent of their calories from animals and animal products compared with the 100 per cent commitment of the sharp-toothed prosimians).

We humans have a lot to say about what goes into a pregnancy and with the advent of evolutionary science we have found new ways to harp on old ideas. Long, drawn-out periods of pregnancy seem like a foolish and possibly even dangerous idea, running around the evolutionary savannah of our imaginations, craving pickles and peanut-butter ice cream.* Modern medicine, and people in general, have identified the pregnant woman as one of the most vulnerable members of society – and subject to a great deal of risk. The dangers facing her are so many that it becomes impossible to even quantify them; that ice cream, for instance, could have been unpasteurised. Your salty pickles might kick off hypertension. Running? On a savannah? In the heat? You're as good as dead already. The list of things that pregnant women mustn't or shouldn't do in the industrialised world is an extraordinary example of how risky we consider pregnancy to be. We are hyper-aware of the potential for pregnancies to fail, even if the figure of

---

* The fact that neither peanut butter nor ice cream was readily available prior to the last century has no impact on their current status as 'typical' pregnancy foods in my homeland. Then again, the store is likely to be out of mammoth.

one in three pregnancies ending in miscarriage is still a surprise to many.

As a species, we worry – and when we worry we start fussing. Around the globe, there are plenty of fussy taboos for pregnant women to suffer.* Plenty of places have (or recently had) rules on what you can or can't eat while pregnant – nothing too spicy in Peru and absolutely no to pineapple in Bangladesh. Eating fish will give your baby eyes like a fish in Namibia, while eating fish with teeth is dangerous in Bolivia as those teeth can cut the umbilical cord. You have to be very careful of any animal mother eats, because in Tanzania babies will act like whatever was on the menu. Eating weird-walkers like reptiles and tortoises will screw up how your baby walks in the Central African Republic, snails make them lazy in Nigeria and eating a frog in China will make your baby prone to misbehave. Eating camel in Niger will make your pregnancy last for a year, while all sorts of foods will risk it: mangoes in Nepal and India, papaya in India and Thailand. Meanwhile, cold food is a bad idea in Burkina Faso and sour food a no-no in Vietnam. Caffeinated drinks will make Thai babies stupid, while eggs will make babies in Zambia hairless.

Like all food taboos, there is a fascinating wealth of anthropology behind each and every one of these rules. Many restrict meat or eggs in particular, which you would think was somewhat counter-intuitive because meat is such a good source of nutrition and presumably you want a well-nourished baby. But many human cultures seem to suggest putting the brakes on bacon for the very sensible reason that it is *too* good a source of nutrition. As we saw in

---

* Maintaining a career seems to be the biggest one.

the last chapter, growing too big a baby risks the twin devils of pre-eclampsia and diabetes. In parts of Ghana pregnant women aren't supposed to have milk, in case they – or the baby – gain too much weight and have a difficult delivery. Some taboos seem to have no good scientific basis, however, like the banning of leafy green vegetables in Ethiopia.* Others aren't even food related, like the unfortunate mothers of Guatemala who cast their eyes to the sky at just the wrong time – seeing an eclipse could give the baby a cleft palate.

But what if there was something you could do while pregnant – or conceiving – that would change the kind of baby you had? There is, of course, one variable that everyone seems to be interested in, and evolutionary anthropologists share with much of the human population an interest in biological sex† as the means of transmission of genetic material from generation to generation. Perhaps no primate was as famously, and murderously, consumed with this question as England's notorious King Henry VIII.‡ Henry is accused in popular history of breaking with the Roman Catholic church because they wouldn't allow him to divorce his first wife, one of a series he would marry, after 23 years of an otherwise presumably unobjectionable first marriage failed to produce a male heir. Anne Boleyn, his second wife, similarly failed to produce a son; she lasted three years before being beheaded and replaced by Jane

---

* Or proscription against a celebratory glass of fizz by certain hard-line elements in the medical establishment. You know who you are.
† This discussion is of biological sex, which is tedious and genetically determined. It is not a discussion about gender, which is fun and whatever you damn well want it to be.
‡ One shudders to think what a late-Tudor gender-reveal party would be like.

Seymour, who did manage a son but died shortly after.* We now know this to be monstrously unfair. Given that human daughters are usually† made of two X chromosomes and women are only ever able to contribute an X, that fatal second X must have come from Henry's gametes. His wives could not have chosen the biological sex of their child; nothing they could do could change the outcome of the random genetic lottery that determines biological sex and they had no say in the matter. Or did they?

As a baseline, human babies are born at a gender ratio 101:100 of about 51 per cent male to 49 per cent female. There are two chromosomes available to dictate biological sex, so you are generally limited to two choices though by all means not always, and those two options appear to be distributed sufficiently randomly that most sexually reproducing animals have an equal ratio of males to females. It is when you start getting skewed ratios that biologists get excited. If the sex ratio at birth can be skewed, then it follows that something must skew it. Clearly there are some behavioural explanations for why you might find animals in different ratios in the wild; for instance, male African lions are famously disinclined to support their brothers in adulthood and only a few will

---

* As many of Henry VIII's wives died of childbirth (Jane Seymour, Catherine Parr – though not with Henry's child) as were executed (Anne Boleyn, Catherine Howard), which begs several questions about the adaptive value of encouraging little girls to aspire to be princesses.

† Technically there are a variety of chromosomal combinations that can result in a baby with female genitalia, including a single X (Turner's Syndrome), and XY (Swyer Syndrome); some males might also carry an extra X (XXY, Klinefelter Syndrome). Biology *hates* being told to colour within the lines. Also, the chromosomes that can make up a female *gendered* child are wide open.

survive to rule the jungle.* This is an example of a survivorship bias, however. Plenty of boy cubs are born – they just don't make it to adulthood.† This has plenty of impact on the lion mating system, leading to the well-known pride structure where a boy lion sits around looking imperious while his harem of females goes about doing all the work. Primates experience the same thing – and often the males aren't even dead. Our macho silverback gorillas may have a harem, but for every one-man band there is a solitary dude just hanging in the jungle, waiting on their chance to fight for a group of their own.‡

What about controlling the numbers of males and females that are actually born? Is that even possible? Here we can look to primates again. Studies of the adorable mouse lemur have shown that not all mouse lemur litters are a random assortment of males and females. In their native Madagascar, mouse lemur females stay near to the place they were born, while males scatter more widely. A surfeit of daughters might mean increased competition over resources in the home territory, a situation unlikely to benefit the mother or the offspring. Males, however, can be birthed safely in the knowledge that they will shortly be someone else's problem. A simple experiment compared the sex ratios of mouse lemurs born to mothers that were housed with a handful of other mouse lemurs to those that

---

* In *The Lion King* in real life, Scar would 100 per cent have killed Simba. Also, Simba would have eventually taken three to four more wives, and spent his life fighting off rivals. If Nala ever kicked him out he'd either roam alone or fight for a new pack and kill all the cubs that weren't his. Nature is mean.
† And if they do, some absolute asshat goes and shoots them dead for *fun*. And Freud. Humans are awful.
‡ Some of these loners even form the gorilla equivalent of boy bands.

had been isolated. The crowded mothers produced sons at a rate of 67 per cent, while those left in peace produced daughters at a rate of just over 60 per cent – a significant difference and an excellent adaptive strategy to what would be challenging conditions in the wild.

What about humans? Was there really something Henry's wives could have done to save themselves?* There is evidence that, in certain circumstances, human sex ratios at birth do change. One of the most famous examples in evolutionary anthropology is exemplified by skews produced by differences in social status among the parents. The seemingly inescapable Trivers, along with Dan Willard, is again responsible for the theory underpinning this in the form of the Trivers–Willard hypothesis. This hypothesis expects that mothers with the means to invest in their children can afford to have the kind of baby with a high-risk strategy towards reproduction. It is only adaptation if one sex has more variable success in reproducing – and for primates, including humans, that would be males. Daughters almost always reproduce and their output, as it were, is fairly steady. Both high- and low-status daughters are still likely to have some offspring, so having them is a safe bet. Sons, meanwhile, are far more of a gamble. Some very successful sons might go on to have leagues of offspring, while the less-successful might have very few or even none.

From a genetic point of view, it makes adaptive sense for mothers who might not be able to invest in their sons and ensure reproductive success to opt for daughters, whereas a higher-status mother, with all the social capital in the world and the prime nutrition status that comes with it,

---

* Aside from the fairly obvious advice not to marry.

might be able to boost her legacy through having sons. This hypothesis has led to some interesting field studies, including in humans – one such even found that billionaires were more likely to have sons, by about 10 per cent. However, about half of the studies came up with little or no support for the Trivers–Willard hypothesis and there are things happening on a global scale that can't be attributed strictly to status differences – in the Western world, for instance, the number of daughters has been creeping up for several decades, something that might correspond with the worldwide increase in fertility problems. It might be that sex ratios do reflect some sort of adaptation, but simply looking to the status of the mother (even if it's via her mega-rich partner) is insufficient to capture what is really going on.

And what is really going on? It's all very well to say that mothers 'choose' the sex of their offspring, but there has to be a mechanism for this – one other than prayer, eating a particular fruit,* or any of the other utterly unproven nonsense that appears in broad circulation. There are, however, mechanisms that, while they may not be under conscious control, do seem to affect the otherwise random allotment of male and female gametes. Here we have an incredibly wide array of 'possible' influences. Males might produce more X- or Y-carrying sperm; this does seem to vary in individuals and even within the same individual depending on how long they've been abstinent (less time

---

* One can only applaud the efforts of the editor of *Mother & Baby* magazine, who chose to herald the absolutely imaginary discovery of boy-conceiving powers in the banana with the phrase 'this divisive fruit'. I imagine this was a complicated but worthwhile lead-up to a terrific banana split joke that sadly has been lost to the sheer inanity of the subject.

since sex, more Y-chromosome sperm). Somewhere in the fertilisation process, X or Y sperm might win or lose out more often – this might be linked to paternal age, maternal age or other processes regulated by the heady mix of hormones that swirl around conception. Alternately, it might be that boy embryos or girl embryos are hardier and less prone to genetic errors that cause miscarriage. This has been argued to be the case for X-linked conditions, certainly – a male embryo with a bad X is stuck with a bad X, but a female has another X to potentially work with.

One aspect of fertilisation that has received considerable attention is the timing. Moving conception either side of the few-days ovulation window available seems to influence sex, with more boys being born from conceptions early on in this window and girls from slightly later; one wonders if this is the basis for the superstition in ancient Chinese medicine that one conceives girls on odd days and boys on even days as counted from the menstrual period.

Our fixation on gestation is tempered by the very real value we place on our offspring – the same impetus that leads us to invest so much time in birthing and raising them. You can see a little of that in the lack of clear-cut clinical evidence for some of even the most common recommendations for pregnant women. The National Health Service in the UK and the USA's Centers for Disease Control and Prevention, for instance, both insist that no level of alcohol is safe for women, because there is no clinical basis for establishing a minimum safe level;[*] despite this, women around the globe will tell you to mind your own damn business when it comes to a glass of wine now

---

[*] This is because 1. Humans can't remember what they consume; 2. It is not considered ethical to get a bunch of pregnant ladies drunk and 'see what happens'.

and then. Women actually can and do run in pregnancy,[*] which is perhaps unsurprising given that being an immobile blob with a hand swollen from hypertension stuck in a pickle jar is unlikely to have been an adaptive strategy when we still had cave lions around. When we really start to see the consequences of our choices – how and when we physically invest in those babies – is in what comes next: actually having them. Because not only is birth a pretty clear pinch point in the evolution of our species and its childhood, it is also an absolute mess and it's a wonder anyone survives it at all.

---

[*] I ran up until 38 weeks, if you allow a *very* loose definition of the verb 'to run'. The ligament-stretching hormone relaxin that is supposed to ease open your pelvis to let the baby pass in the final stages of pregnancy also works on ankles, so if you don't fancy going down like the Hindenberg in front of other joggers it's pretty sensible to stick to exercise your ligaments can handle.

## CHAPTER SEVEN

# Cackle, Cackle, Mother Goose: Having a Baby

Cackle, cackle, Mother Goose,
Have you any feathers loose?

Evolution is always something that happens to other people – right up until you try to have a baby and then, wham, it hits you like a 4kg bowling ball right in the genitals.

The thing with human babies is the sheer horror of having them[*] – and the fact that our babies face a pretty big risk just being born. This has to be considered as part of our overall childhood investment strategy too. If we are thinking about growing kiddos as a series of life-history and investment decisions, let's not forget we actually have to survive each stage. Human birth is risky, painful and something that we are just not very good at. Considering that reproduction is *the* major hurdle a species is required to leap to survive, it is absolutely extraordinary that we even exist when 'natural' – for which read, without access to medical assistance – mortality for women giving birth is over one in a hundred. More than one in every hundred women will die giving birth. And that's not a one-time thing; the risk repeats with every birth. So, if you have one baby, you will have rolled the dice once, but continue on

---

[*] *The Diagnostic and Statistical Manual of Mental Disorders*, or *DSM* as it's usually known, defines 'tokophobia' as 'the unreasonable fear of giving birth'. Given actual statistics of death in childbirth, this definition hardly seems … reasonable.

to what is probably peak fertility for our species with huge energy stores from a nice agricultural lifestyle behind you and you are essentially rolling the dice every year or two. Those are big dice, sure, but still, it's an incredible risk – you wouldn't buy a car that had a one in 100 chance of the brakes going out each year would you?

Apparently, yes you would. As many have noted, there are a vast number of people alive on the planet and the majority of them will have arrived through the standard process of human birth. There is hardly anything standard about it, however. If anything, human birth can be described as convoluted, protracted and easily complicated. These ridiculous impediments to life are, however, deeply tied into our evolutionary history, not just in bone and tissue, but in our social being – the very way we hold our human societies together. But it begins in blood.

In our species, birth is notable primarily for how awkward we find it – other animals manage to have babies without making quite such a song and dance about it.* While no one is likely to top the poor hyena in terms of birth horror stories – giving birth through a penis-shaped clitoris to a baby with a full set of teeth is a high-water mark no one wants to surpass – we are not entirely alone in our travails. But wait, you say, dimly recalling your high school biology class. The reason human babies are such a pain is so obviously an evolutionary adaptation that we teach it to those self-same babies not too many years after they are born. The main feature of hominid evolution that nearly everyone will have been taught† is that our ancient ancestors learned

---

* Or turning it into a series of anecdotes for a book.
† Rather depressingly, the teaching of evolution has taken a bit of a nosedive in certain locales, because nothing threatens social order like considering the world over the long term.

to walk upright, changing the shapes of their pelvis as they went along to accommodate the new savannah swagger. At the same time, brains were expanding, perhaps because of all the new foods we could efficiently walk to, and this created what anthropologist Sherwood Washburn coined the 'obstetric dilemma'. We have awkward babies because we have carefully balanced the demands of our unique upright posture – bipedalism – with our enormous human brains, the great thinking machines that have allowed us to dominate the planet and maybe even eventually the stars. We are special and so are our births. So far, so good – except no one told the squirrel monkeys.

Squirrel monkeys, it turns out, are also terrible at having babies. They are not known for their planet-dominating brains, they struggle to walk upright more than a few paces and yet they also suffer difficult births. Fifty per cent of the babies born to one captive group didn't make it through the birth process – that's way worse than the human rate of two per cent, though humans can do worse if you start adding all the health burdens of inequality and disease. But the squirrel monkey's existential struggle serves to bring our own further into question: what the hell is wrong with us that we have such high fatalities right at the moment of reproduction and how did we come to such a pretty pass to begin with? And whose fault is it, if it isn't our habit of walking upright?

The mechanics of human birth at least are simple, well established and utterly contrary to sense.* One takes a

---

* I would like to take a moment to direct the reader's attention to the linguistic anomaly that means that speakers of North American English *get* pregnant, whereas those from the UK somehow mysteriously *fall* pregnant. One wonders at the mechanism behind this. Pit traps?

mammal, grows a baby to nearly* the same size as the
mother's pelvis, and then forces a complicated solution to
the resulting sofa-in-a-stairwell issue through three stages.†
For reference, there is the first stage of labour, when
contractions start, which can go on for hours; then there is
the rather unpleasant second stage, which is very huffing
and puffing like you've seen on TV, but at least is mercifully
measured in minutes and results in a baby; and then there
is the third stage, which is actually just getting rid of the
placenta. It was quite the vogue in the mid twentieth
century to describe uterine forces with barometric
measures, in terms of exactly how far a contraction would
move a millimetre of mercury, so you can think of the
whole process in terms of pressure, just as you would
something more pleasant, like the toxic atmosphere of
Venus. As an example, the atmosphere of Earth exerts the
same pressure as about 760mm of mercury (mm Hg) and
the pressure on the uterus in labour is measured‡ as rising
about 10mm Hg on average, going up to 50mm Hg at
maximum levels.§ This sounds absolutely negligible, until
you realise that by the time uterine contractions are getting
to the strength they will actually expel a baby, they measure

---

* It is this 'nearly' that saves us. Just about.
† This is not terribly interesting unless it is happening to you, in
which case it certainly captures the attention but you rather wish it
wouldn't.
‡ They are measured on a tocodynamometer, which is the funny
little thing they strap to your belly when you are in labour and feeds
out a steady seismograph of contractions. The hospital will then give
you this readout, whether you want it or not, or whether you have
remembered to bring to the hospital a folder or indeed plastic bag
big enough for 36 continuous hours of feedback.
§ The atmosphere of Venus is around 70,000mm Hg, if you really
want to contemplate it.

more like 2,000mm Hg. That's the same force as two and three quarters of those Earth atmospheres being applied across a very small band of tissue. It is the pressure equivalent of sinking your stomach and *only* your stomach 26m below the surface of the sea, repeatedly, at 30-second intervals.

However you wish to imagine the force involved, this initial process of labour is almost certainly worse. It's not nearly as bad as what's coming next, though. Next is the second stage of labour, and that is what kills us, and those poor squirrel monkeys. This is the stage where the soft tissue cervix of the uterus is as open as it's going to get, and the baby's head starts to move out and down into the pelvis. In the vast majority of mammals, this is not a big deal. The baby, or babies, travel down the pelvic canal and out into the world with some powerful contractions, but very rarely are there any actual issues with exits, unless the animal has been overbred by humans. It has been theorised that the hormone relaxin, which can loosen up ligaments in the body, builds up in pregnancy so it can unspring the bones of the pelvis a bit to give a bit more room, but this remains unproven.* Either way, the feral cats haunting my doorstep (and evading capture and neutering) reliably produce kittens in the old sheep shed next door with zero intervention, unless you count me chasing off fighting tom cats every so often. Meanwhile, purebred dogs like the Boston terrier or French bulldog, so beloved of celebrities and social media influencers, require a caesarean section in

---

* It is also a handy hormone to blame for when your feet go flat in pregnancy or your joints wobble uncontrollably, rather than the 5–10kg of cheeseburger you have asked them to carry.

nearly 80 per cent of births or they will die.* The cat's pelvis is plenty big for those kittens, while the shape of the dog's pelvis, and their puppies' heads, has been manipulated by selective breeding into something that no longer works. And while it's all well and good to react in horror that we could do this to poor little puppy dogs, there remains the small matter that we seem to have also done it to ourselves.

And here we come to the crux of the issue: the human pelvis versus the human baby head. The human baby head is usually ever so slightly bigger than the passage from which we wish it to exit. The other great apes have not made this mistake – and that's despite many of them chunking up on greater proportions of their adult brain size while in the womb than we do. Actually, we're not talking about all human pelvises. We're talking about female pelvises. The three bones that make up your hip girdle are shaped differently in males and females – one of the few examples of sexual dimorphism we can identify in our species. As you might imagine, the female hips are wider and the underlying bone structure more spread apart to facilitate that baby.† This is one of the foundation stones of the idea that hips are as wide as they are to get a baby out – because we see there are clear differences in females, who might need that capacity, and males, who don't. In humans, we like to blame big hips on the to-the-death battle of birth.

---

* Note to the person who spams my Amazon reviews calling me a dog-hater – I actually quite like dogs, which is why I am against irresponsible breeding. Also, thank you for caring. A lot.
† And the job of the osteoarchaeologist who wishes to ascertain biological sex from a skeleton. This many chapters in you didn't think I made my living writing books, did you?

But those great apes, the ones with the easy births? They also have sexually dimorphic hips. Chimpanzee males and females, who have babies just fine, have the same differences as male and female humans do once you scale everything up to account for our larger body size. Actually, so do my feral cats, and even that possum whose babies are so small they fit in her handbag. New research suggests that mammals just have sexually dimorphic hips, from way back, so maybe that tight fit isn't actually the immediate response to baby size we think it is. More recently, a survey of the size and shape of hips across 24 different human populations found that geography had a bigger effect on our hips than birthing destiny.

So if the problem isn't the shape of the hips, maybe we can blame the baby. Well, on the baby's end, there is clear

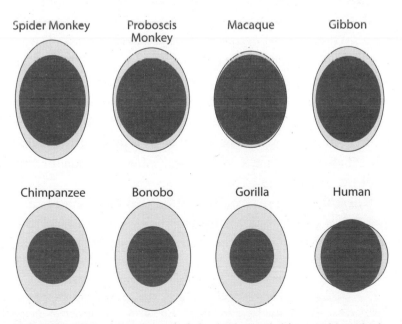

Figure 7.1. *The relative size of the birth canal (outer circle) with the relative size of the newborn skull (inner circle) in several different primates. After Rosenberg & Trevathan (2002).*

accommodation of this ridiculous state of affairs in the bones of the newborn skull. When you are born, the cranial vault, the bones forming the eggshell around your brain, are not only small, they are in pieces. The frontal bone that makes up the solid, beer-can-defeating adult forehead starts off as two pieces; the sides are two different pieces and the back is four. They float over the surface of your thinking parts held together with only collagen, muscle and skin. While this sounds on the face of things – or rather skull of things – to be a terrible idea, it actually allows for much greater flexibility when subjected to all that pressure described above. The newborn head easily deforms and bounces back because it needs to get out and then it needs to do some serious growing. This is why we have a 'soft spot' or fontanelle rather than a full bony helmet at birth, and our malleable, semi-cartilaginous, not-quite-finished skull bones are also why if you leave your baby lying around on hard enough surfaces long enough – or, in some cultures, deliberately bind up their heads – you can actually change the shape of the baby's skull.* For instance, the flat back of the head that modern clinical practice uses as an indication of neglect was once the height of dedication among conscientious Mesoamerican parents who sought to give their babies the best possible skull shape for health and spiritual well-being.

There is a second accommodation that our babies make, a solution to the ridiculously tight fit between head and pelvis that for many years researchers thought was unique

---

* This doesn't bother the baby, unless they grow up to be an osteoarchaeologist and have to spend time on the internet shouting at people who think that elongated skulls are aliens, and not indications of social status and cultural affiliation. Hint: it's *never* aliens.

just to humans. As the baby comes down the birth canal, the head turns to present the least possible obstruction to the bony cage of the pelvis. The baby's face isn't towards the mother's front as in most other primates, but instead turns so that the baby emerges facing backwards, and a bit sideways. That last minute twist is what allows us to be born. We used to think that no one else did this, that this was a human-only innovation to the vexed question of getting a big baby out of small hips. But actually, there are several other primates who do a little turn on the pelvic floor: gibbons, macaques and those unfortunate squirrel monkeys. Even chimpanzees, they of the easy births, have babies that turn slightly. This is not the unique trait that we imagined when we discovered it, and this has unfortunate consequences for many of our more interesting theories about birth and human society.

The odd, backward-facing baby of the human animal has for years provided fertile ground for theories about why we give birth like we do. Most primates do what my feral cat does, which is run away and hide when the time comes.

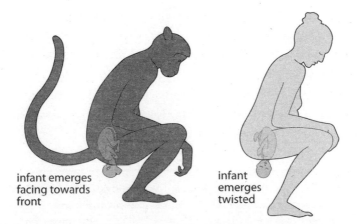

infant emerges facing towards front

infant emerges twisted

Figure 7.2. *The relative position of a newborn macaque (left) and a newborn human at birth. The human has made a 'twist' in the birth canal.*

Even in terribly social and gregarious monkeys, birth seems to be a time when animals would rather be alone. Humans, however, give birth in company.* Attend a large-ish hospital for the event and you will meet no less than a dozen people: nurses, doctors, midwives, lunch ladies, cleaners, all of whom are there to help you usher your new little human into the world. Even in out-of-hospital births, women are usually surrounded by friends, family, experienced older women, midwives or any combination of these. This is really a quite overwhelming group experience compared with the births primatologists have observed in monkeys and apes, and the suggestion has been that our twisted exit strategy has something to do with it.

Monkeys whose babies exit face up have their spine and head in such a position that the mother is able to reach down and assist herself in the process of giving birth. The half-twist of the human, however, means the baby is facing the wrong way to be pulled out from the front, so mama needs some help. That help — the midwives, doctors, friends and relations — is one of the things researchers have suggested drove our social evolution. Humans, researchers suggested, need the kind of friends you can give birth with — and that is an evolutionary step towards the big, interdependent societies that have made us who we are. It takes a very special kind of social bond† to let someone anywhere near you at your most vulnerable, after all, and it is humanity's unique emphasis on sociability that allows

---

* This is true even in the middle of a pandemic. They might throw your much-reduced, one-person social network out of the building less than an hour after the birth, but they introduce you to your new best friend co-codamol shortly after. Also, anaesthetists are very chatty.
† And/or fentanyl.

us to have these unreasonably sized babies in such a convoluted way.

Except, except, except. There *are* monkey midwives. Langurs, spindly little monkeys found on the Indian sub-continent that are approximately 200 per cent tail, have been observed helping out during births, though admittedly this has only been observed once. But the chimpanzee, who we discover also has a bit of a wending path out of the womb, much like a human? She *isn't* getting any help, so we can't say that midwives, or the sort of social bonds that let you deliver someone else's baby, are the exclusive result of a twisted birth process. Primates are almost all born head first, but only a few twist around. Yet when we see it, it doesn't necessarily accompany a tighter fit between baby and pelvis. That chimpanzee will get away with a much easier birth because she has plenty of hip room, but her baby's head will still rotate the same way a human baby's does trying to escape the narrow confines of our awkward human pelvises.

The other thing that chimpanzees will get away with is, of course, not being bipedal. Chimps can get around upright, and they can walk on two legs if so inclined, but when they do they stagger around the place like a sailor on shore leave. Their legs have to sort of swing out to the side with every step, because that is how you attach a chimpanzee-shaped short-necked femur to a chimpanzee-shaped hip. We, on the other hand, walk upright as a matter of course and, while capable of doing a great number of very silly walks indeed, are actually fairly efficient at it because our femurs have long necks so we can achieve a knock-kneed stance for optimal striding. Blaming the shape of our hips, which are very appropriately shaped for walking – they are more squished between front and back, and wider side to side than the chimp's – for our difficulties in birth is perhaps the best-known and longest-standing evolutionary hypothesis we have for our species. The final half-pike in the birth canal was for a long time key evidence in the argument that we had forced evolution to literally take a detour in order to accommodate our big brains and upright posture. But since now we know a chimp baby turns too, we can start to question the orthodoxy of the stand-and-deliver hypothesis.

If it's not the shape of us, then we might want to start thinking about size. This is where the size of the baby starts to really, really matter. We certainly have not left much room for error. But we are not the only idiots on the birthing block, which is either reassuring or depressing, depending on how you feel about big babies. Kiwi birds, for instance, lay eggs a quarter of the size of their total mass, which is way out of proportion − six times larger − than you would expect from a bird their size.[*] Humans? We are churning out babies that are big, big, big for the size of their mamas. Not kiwi-sized, for which we should be thankful, but we are still a good 6–8 per cent the size of our mother when we are born. Chimps, meanwhile, are only half that − 3 per cent of their mother's size at birth. In a cuddly, pinchable, round-cheeked kind of way, our babies are born with the advantage of extra ounces of free energy, all courtesy of mother. About 15 per cent of a human baby is pure lard. Compare that with the 3 per cent fat content of our closest living relatives the chimpanzees and you start to see what a dramatic commitment we have made to fat storage as a way of life. We are big at birth − and we get there fast.

---

[*] You can see from their faces that they know this.

Most of the monkeys and apes follow a more leisurely course to growth than their similar-sized animal counterparts. The mouse lemur, for instance, tops the scales at around 85g while our old friend Pizza Rat can triple that, easily. Yet a mouse lemur pregnancy will last around 60 days, while the Pizza Rat only needs a third of that time. Three times the size and three times as fast – there is a clear difference in the amount of investment our Pizza Rat is putting into her pregnancy. Where the mouse lemur is taking her time, Pizza Rat is hurtling towards the end of a short but intense pregnancy. Well, fine – those little baby rats are even more unfinished than primate babies, coming out into the world furless and with eyes and ears shut. Pizza Rat can sort out transferring the energy to grow those rats after they're born. Adjustment made.

But, and this is important, what then to make of Prince Demidoff's galago?* This bushbaby species, another primate, is the same size and weight as a mouse lemur, but takes almost four months to gestate compared with the mouse lemur's two months. Or, indeed, the only ape we refuse to call great (most likely because they are actually quite small, around 5–7kg), the gibbon, which needs nearly seven months to grow a 300–500g baby. The similarly sized macaque, meanwhile, has got the whole business over and done with in an economical five and a half months. Moving closer to our particular branch of the evolutionary tree, the great apes have babies that weigh in quite small – our chimpanzee mama, who will top out at around 40kg, has a

---

* Disappointingly the species was so-named because it was identified at what is now the Vernadsky State Geological Museum under the patronage of keen naturalist Pavel Grigoryevich Demidov, and not because the aforementioned literal bigwig Russian royal spent the 1790s tromping around Madagascar naming primates.

baby that's a little less than 2kg, about 5 per cent of her size, and an even smaller percentage of the dad's 50kg. Gorillas, big beasts that they are, are also born at about the 3 per cent mark, with tiny 2kg babies born to 70–110kg mothers (and fathers up to nearly 230kg). So when we show up at 6–8 per cent, we know there is something else going on. In Chapter 5 we saw the value in fattening up mothers and potential mothers against the risk of an uncertain (or certain to be bad) food supply. But there is another use for all that extra energy and it's one that particularly interests evolutionary anthropologists.

In 1995, anthropologists Leslie Aiello and Peter Wheeler published a theory arguing that fat is critical to building the body parts we need to become human. The 'expensive tissue hypothesis' was an argument that our impressive capacity for fat storage is in fact adaptive; we use that stored fat to budget for the running costs of tissue that we couldn't otherwise afford to run: our big fat massive brains. According to the authors, a human brain accounts for an impressive 16 per cent of a human's basic metabolic rate. To add more to the brain-growing budget, we have to take away from somewhere else. Aiello and Wheeler proposed that one of the areas that suffered was our gut, another expensive organ, and one that could take what a management consultant would recognise as 'significant restructuring' if you were to make sure you only asked it to deal with high quality,* nutrient-rich foods. A richer diet in our ancestors' past – perhaps as far back as a few million years ago, with the earliest *Homo* species – would have

---

* High quality in an evolutionary sense means exactly the opposite of what it does now – forget dense, fibrous plants that take ages to chew and digest, we're talking bacon on your cupcake. If either of those things had existed 3–4 million years ago.

allowed for efficiency savings in the gut to be transferred straight to our hungry brains. Further testing of this hypothesis in a much wider variety of species showed that not all animals that have big fat brains have a reduced gut, but there is an association with how much fat an animal can store in its body. Alternative energy savings have been proposed, for instance in our neat energy-conserving locomotion style, but they seem to be offset by the amount of it we do. Whichever mechanism is at work, the end result is clear in the production of human infants with fat, expensive brains.

Primate brains take quite a lot of effort – and calorific investment – to build. Take the bug-eyed, swivel-headed, lizard-crunching tarsier. Tarsiers are mad little halfway houses between the lemurs, lorises and aye-ayes that make up the prosimians, and the more derived (or, changed-from-the-last-common-ancestor) other monkeys and apes. Their eyeballs – each one by itself, not even together – weigh more than their brains and are quite literally bigger than their stomachs.* This indicates a rather concentrated evolutionary adaptation towards eyeballs, which comes as no surprise if you realise that tarsiers are the only primates that survive almost entirely on an all-protein diet. They need those eyes. If they can't leap after lizards or other protein sources with a degree of accuracy, they aren't going to do very well. Thus, we see that tarsier mothers spend ages gestating their babies, so that the infant tarsier is born with a proportionately (for its tiny size) larger brain mostly devoted to operating those eyeballs.

---

* Which suggests immediate tarsier ancestry in my genetic line, according to the Depression-era-born grandmother tasked with feeding me lunch as a child.

Pound for pound tarsiers work far harder at pregnancy than even the meatiest of humans. A female tarsier will build a baby tarsier that weighs about 19 per cent of her own body weight in 170 days of pregnancy, averaging about 0.17g of additional tarsier a day until there is a whole 30g-worth of newborn. Humans, on the other hand, will drag their feet building something only around 6–8 per cent of the mother's weight through 267 days of pregnancy, averaging 12g a day until we get those 3.3kg babies. It's not the same level of effort – 0.17g to a tarsier is about 0.08 per cent of body weight, while 12g is 0.03 per cent of even the tiniest human mothers.*

Researchers have estimated the likely calorie cost of making such clever things as apes and there are definite trade-offs. We saw previously that gibbons, the not-so-great apes, despite going around the world looking like they've recently stuck a finger on one of their long, long arms into a plug socket, have longer pregnancies than similar-sized macaques. However, macaques are born with about 60g of brain that they need to bring up to about 100g when they reach adulthood, a trek of 40g, so they are around 76 per cent done with brain growth by the time they are born. Gibbons start off with around 55g that they need to build up to around 110g, so they need to double their at-birth brain weight and are only halfway there – 50 per cent – as newborns. This means that in utero gibbons actually produce less brain tissue, over a

---

* These statistics are calculated using a classic comparative paper from 1985 by Harvey and Clutton-Brock on 'Primate life history variation', which suggests the average human female weight is 40kg, or around 90lb. It is my personal experience that 90lb is not what human mothers weigh. I last weighed 90lb when Nirvana were still playing shows.

longer period of time, and they make it all up after birth.
A rhesus macaque, by contrast, had a brain 'growth spurt'
while still in the womb, about two-thirds of the way
cooked.

So why do some animals work harder at pregnancy than
others? There are two ways to funnel brain-growing
calories to a baby: put the effort in while the baby is still in
utero, or do it after birth. Chimpanzees, for instance, are
born with brains about 40 per cent of adult size. Us? A
mere 30 per cent. The choice of when and how to make
those investments are what shapes childhood. To get our
big-brained adults we need to provide enough energy to
our kids so they can grow not just the usual mammal suite
of limbs, *etc.*, but this enormous central command edifice

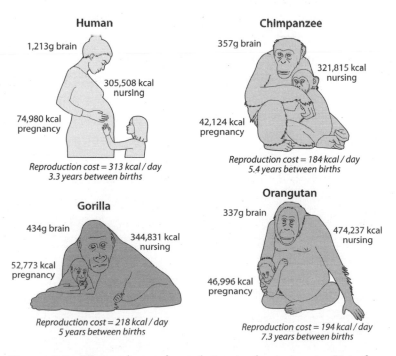

Figure 7.3. *The total cost of reproducing in the great apes. Data from
Pontzer* et al. *(2016).*

that has gone way, way over the traditional mammal budget. When you are born to run, or at least stumble around, like a baby giraffe, you want a brain that is pretty much sorted: precocial animals tend to have their big brain growth inside the womb. For the animals that are going to have to deal with the mewling and the neediness for some time after birth, you might as well leave the brain growth until later, when you can start dragging pizza into the equation. Us? Yep, we chose pizza.

If we actually wanted to have brains as large as a chimpanzee's when we are born, proportionately, we would have to be pregnant for 18–21 months.* Instead of building big-brained newborns, however, we're building big-brained babies outside the womb, without even the benefit of a pouch. That is the thing we can see as one of our lineages' greatest adaptations: procrastination. We really concentrate our baby brain growth in the 90 days immediately after birth, letting our babies try to muddle through a brand-new world while expanding their brains at a rapid clip of 1 per cent a day. Us terribly clever apes have looked at the energy it takes to grow a baby, a really smart baby, and decided to spread the investment out as long as we possibly can.

Our complex energy needs – getting to a reproductive state in the first place, the crushing weight of our giant intellects or whatever else is filling the space between our ears – require energy. Recent work has finally managed to identify another nifty adaptation that humans have made, relative to the other remaining great apes: running hot. Despite what the middle-aged reader of this book might suspect, the human metabolism is in fact much faster than

---

* In my immediate unscientific opinion, this is too long. Happily, evolution backs me up on this one.

our ape counterparts. A 40kg female chimpanzee burns through a little less than 1,500 calories a day, while the 60kg-ish female human blazes through more than 2,000 and change. That's accommodating all those expensive organs and leaving room for our crowning achievement – holding on to extra calories so we can have more babies. When our rather maniacal placental delivery system starts asserting the demands of the baby in no uncertain terms then we can start to see where some of that extra energy capacity is going – we are investing it, like it or not, in our brainy babies. But even our jacked-up metabolisms aren't enough to build the babies we need entirely out of our own bodies, so out they come, as fat as we can get them, early, helpless and protesting into the baffling world of human life.

## CHAPTER EIGHT

# See-Saw, Margery Daw: Cultural Adaptations to Birth

See-saw, Margery Daw,
Sold her bed and lay on the straw;
Sold her bed and lay upon hay
And pisky came and carried her away.

Somewhere along the line of evolutionary history that leads to us, we decided to go big, and the one-two punch of maternal energetics (running hot) and baby size (fat and cute) came up against the bony reality of our pelves. There is a fair amount of guesswork involved in working out where and when transferring fat and fatty brain potential appeared in our species, but it is an investment that we have been doubling down on ever since. Squeezing mama's metabolism to the point of near death in some cases to grow these fat babies that we then have immense trouble birthing is a tightrope we have been walking probably as long as we have been walking; the origins of our troubled births probably stretch well back before our species appeared. Lucy, the *Australopithecus afarensis* way back in our hominid ancestry would have loved an epidural. Earlier australopithecines like *A. sediba*, however, don't seem to have been built for the twist down the birth canal. This really does beg the question – if we have had trouble having kids for millions of years, how the hell have we survived? The answer, as much as we can ever answer any question about our abiding weirdness, lies in the vehicle for investment that has made the most difference to human lives: other people. That one solitary langur midwife has not a patch on the vast cultural resources that surround human childbirth. We have built up edifice upon edifice for helping women have babies, precisely because we are so very terrible at it. So, while midwives may not be the direct result of our funny, twisted passage into the world, they are still a very important part of human adaptation.

If only all of birthing-related culture was so beneficial. Human society has a great deal to say about how you ought to give birth and often what isn't proscriptive is just mean. This is remarkable given that there is not a huge amount of choice in the matter and it is unlikely that you will be able to speak to the manager about any of it in time to do any good. What cultures have to say about birth, however, is often very, very different from one place to the next – and even from one generation to the next. For instance, there are strong ideas about how one ought to have a baby in my natal culture (the US) that utterly clash with those in my adopted one (the UK) – which, by the way, shares a language and all but the last 200-odd years of history with the land of my birth.* I have never, ever, ever wanted to have a baby in the comfort of my own home, even if advocates of a less-medicalised 'natural' birth assure me that it will make me more relaxed and possibly orgasmic during the process.† However, this was an option that was extremely important to the small cohort of women I met who were due around the same time I was, and there was genuine disappointment that home births, in special portable

---

* Worth noting that if you ask an American if they want free medical *anything* they will say yes. This also applies to birth.
† This is, apparently, a thing.

Jacuzzis or not, were disallowed for reasons of a global pandemic. The distance between cultural expectations of birth was further compounded by generation, which I realised while trying to explain to my mother what a birthing pool was and why you would put one in your living room.

Around the world, cultures are understandably keen to invest in the next generation by assisting women with successful births. This leads to the litany of food taboos we saw previously, but also to physical traditions of birth that can vary just as widely as that of hospital and of home birthing pool.* A survey by anthropologist Wenda Trevathan of the actual physical position in which women are encouraged to give birth across 159 cultures shows that sitting is the most common, followed by kneeling and then squatting.† While it is unclear what would have been the case in the past, in many cultures women do not have the muscle strength to squat through labour unaided, so lots of these positions can be aided by clever inventions like the birthing stool, which is a chair with a hole in the seat, or the simple expedient of leaning on things and people. Position, however, is perhaps the least contentious of cultural differences. There are strong proscriptions on women's behaviour around the world as well.

The culture of childbirth across time and geography is utterly fascinating and really ought to be considered in far

---

* What really sold the hospital birth, aside from the free drugs, was the information sheet on birthing pools that indicated one must provide their own sieve.

† Lying down, which is typical in Western medical practice, is only marginally more popular than hammocks, which I can only assume is down to a scarcity of hammocks.

greater detail than I am able to fit in this increasingly oversized manuscript. In the modern day, birth is idealised in different ways around the world. Rather infamously in Scientology, but also in many other cultures, women are encouraged not to make too much noise and instead have a kind of silent birth; by contrast, popular culture in the Anglophone world tends to make a joke out of the quality and quantity of cursing attendant on the process. While the idea that birth is going to be uncomfortable seems near universal, in the patchouli-scented corners of the internet, women are informed that birth can be orgasmic if only they do it right. The underlying principle in both of these cases is actually the understandable goal of achieving a birth that is less traumatic, both for mother and baby, by creating a calm and peaceful atmosphere. However, even the concept of an 'ideal' birth can be used to oppress and subsequently *de*press those women who failed to bring the correct-scented (flameless) candle to the delivery suite, or didn't try hard enough to give birth to their baby underwater accompanied by whale song.*

The long history of interventions and fussing around our births, however, is an unassailable testament to how difficult we have made it. Certainly, knowledge on how to assist birth existed well before writing was even a glint in an exchange token system's eye, but medical history does have quite a bit to say about it. Texts from as far back as texts exist in Ancient Mesopotamia include prayers and rituals (mostly belly rubs) for safe delivery, lamentations for women who have died and even practical warnings

---

* Note to non-UK readers: birthing pools are a thing and expectant mothers are given to understand that the ideal birth takes place at home in essentially an indoor rental jacuzzi. They don't tell you about the sieve until it's too late.

that women with big heads are doomed to difficult labour. Going back some 4,000 years, Egypt has provided reams of medical texts, but it is surprisingly the mythical and magical texts that tell us most about birth.* What may be the world's first purpose-built labour ward – albeit only for deities – can be found in the small chapels proto-Egyptologist Jean-François Champollion called 'mammisi'. These chapels are decorated with birth scenes with midwives fore and aft of the expectant mother, showing either the use of a small seat or a kneeling position. What seems to have been more common in the real world, however, was the use of the magical 'birth brick' or *meskhenet*. These special decorated bricks served as both a surface to labour against and on which to eventually lay the baby for the cutting of the umbilical cord. Perhaps an unexpected bit of obstetric furniture – but then, so is a jacuzzi. The 3,000-year-old Ayurvedic medical tradition of India also emphasizes the importance of midwives (*dai*) and massage. Meanwhile, in early Imperial China, 2,000 years ago, medical texts had a series of proscriptions such as forbidding hair-washing for the whole month before the birth, which when it occurred would see the expectant mother enter a birthing tent – either outdoors or indoors – with just a small number of assistants.

You will note, of course that these are not texts by birth attendants themselves. These are literary texts for a literary elite, who by and large were men and not midwives, the precursor to the male obstetricians who dominated the past few centuries of medical childbirth. Women practitioners faced rather an uphill battle against the social expectations

---

* And possibly abortion; the goddess Bubastis seems to have chosen not to have a child after being raped by the god Seth.

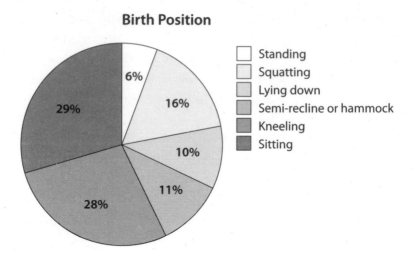

Figure 8.1. *Birth positions around the world. Data from Rosenberg & Trevathan (2002).*

of the day – and that seems to have been true for almost all days since the dawn of history. On one hand, midwives are a human habit far before history even starts. On the other, we are a living species and not a fossil – traditions change, technologies improve. The story of how birth became a matter for medics and not midwives is intimately tied up with larger cultural changes that are still rippling through our societies today, sparking countless arguments on internet forums for pregnant women about how 'best' to give birth.

In considering Western tradition, the case for 'natural' birth was not only medical, but moral. There is a deep history in Christian theology of equating the pain of original sin – the sin of Eve who ate an apple God was saving for later – with the pain of birth and insisting that women must feel the latter to atone for the former. Souls were on the line and examples had to be made – like poor old Agnes Sampson. Agnes, a midwife and general 'cunning woman', was burned alive on Edinburgh's Castle Hill in 1591 for – among other things – 'assisting' the labour of one Euphemia Maclean with a mystery powder, a stone with a hole, and the finger, toe and knee joints of a recently disinterred corpse. While unlikely to have been truly efficacious pain relief, it was not just the potential shamanism of Agnes' treatment – the case against her very clearly emphasises the intention to relieve pain as a condemning factor. Women in this culture were supposed to rely on religious relics and prayer for pain relief, and nothing else. These could take actual physical form like a holy 'birthing girdle' – a sort of maternity sash – which was to be lent out to women in labour like a library book. One of these was recently analysed scientifically and discovered to be liberally coated with the kinds of cells you find floating around during the birth process, suggesting that they really were used. Astonishingly, it was only in the late nineteenth and early twentieth centuries that pain relief during birth was widely accepted in some parts of the world.

One of the best ways to understand the medicalisation of birth is to consider the rise of obstetrics in Scotland from women's work to prestigious and royally recognised male employment. Scotland in the industrial age suffered combined tragedies of a lack of sunshine during the height of industrial pollution and gross inequalities that limited quality food. Many Scottish cities saw cases of vitamin D deficiency in childhood – rickets – skyrocket in the nineteenth century, because the precursor to the vitamin that hardens bones into their proper shape wasn't available from either the sky or the food. This caused an epidemic of rickets, which most people think of in terms of bowed legs, but the same lack of mineral density that causes femurs to deform under the weight of the body also hits the pelvis. That's fine, if you're a boy and do not intend to pass a baby through it. If you're a girl, and you later get pregnant, it's a death sentence.

The nascent discipline of obstetrics was spurred forward by the large numbers of such complicated births. That the chainsaw is actually a scaled-up version of the pubic bone saw – which was invented (twice) by canny Scots for assisting birth as rapidly as possible in order not to lose both mother and baby – is really all you need to know. The first use of a modern anaesthetic in labour was in a woman who had a 'contracted pelvis' from childhood rickets, who successfully delivered while knocked out on ether by the famed obstetrician John Simpson of Edinburgh in 1847. However, he came up hard against the Book of Genesis and its imprecation that 'in sorrow shalt thou bring forth children'. Religious objections remained a major obstacle to pain relief in labour only mildly mollified by the fact that Queen Victoria herself had Simpson put her under for the birth of her eighth child.

Royalty or not, ether and chloroform were rapidly introduced to obstetric practice, allowing interventions that previously couldn't be tolerated, including the caesarean section, itself of considerable antiquity (the clue is in the name). A caesarean section where the mother and the baby survived was a somewhat radical departure from the medieval practice, where infants were removed from the womb after maternal death in order to be baptised, but not expected to survive. Prior to the invention of both anaesthetic and Joseph Lister's discovery of antiseptics, a birth that failed to progress was almost certain to mean death for the mother and the baby. Only in a rare few cases did either party survive; there is a rather shocked account in the *British Medical Journal* from 1974 recounting that Mary Donally, an 'illiterate Irish woman',* saved the life of

---

* It is unclear which of these the educated, male, British author finds more shocking.

a mother whose delivery had failed by performing a caesarean section in 1738 with a straight razor. It's not clear what proportion of births in the beginning of the twentieth century occurred by caesarean section but it does seem to have become something of a fad – and a dangerous one at that, with unskilled obstetricians leaping into surgery. By the middle of the twentieth century, however, the 'C-section' had become a relatively safe and increasingly common procedure, so much so that by the time of writing this the average national caesarean section rate around the world is around 20 per cent of births, even though the WHO estimated (in 1985) only about 10–15 per cent of births required them. Some countries go much, much higher: more than 50 per cent of mothers in Brazil, Chile, Cyprus and Egypt have C-sections, while the figure is 58 per cent in the Dominican Republic.

While there are very, very good reasons to have a caesarean section, like not dying, the elevated rates present in many countries point to real failings in supporting maternal and foetal health leading to serious medical intervention. The joy taken in shaming women who have surgical interventions in their births for contributing to these figures rather overlooks the fact that the factors that drive up rates don't occur in a vacuum – they occur in a society. The nutritional inequalities that underlie conditions that make too-big babies like gestational diabetes and social and economic pressures that encourage later motherhood* in many countries contribute to problems with birth that can only be rectified through surgery. That said, the medicalisation of the old–as–humankind art of being born (and giving birth) has not been without impact – some

---

* Congratulations anyone who could afford to have a child before the age of 60.

sufficiently terrible that you can understand why chilling out in a birthing pool in your own living room starts to sound like a good idea.

The men-midwives of the eighteenth century, who began to replace the women-midwives of the years previous, brought with them plenty of technical medical innovations, like life-saving and entertaining anaesthesia as discussed above. They also reshaped midwifery from a community activity into a money-making business. The introduction of forceps into labour in the 1600s marked a turning point in medical birth, the scooped ends of the Chamberlain forceps making extraction of a live baby a possibility in many cases where previously both mother and child might have perished. Peter Chamberlain, the inventor, knew he was on to a good thing. He set up a maternity empire, breeding a dynasty of Peter Chamberlains who could all perform the miraculous act of resolving obstructed labour with these forceps – and would never reveal their trade secret. They are reputed to have even introduced the use of a sheet during birth – not to shield the modesty of their patient, but to hide the money-making forceps.

By the seventeenth century forceps were popular enough in England to be mocked in the satirical work *The Life and Opinions of Tristram Shandy, Gentleman*, where the narrator bemoans the flattened bridge of his nose caused by a forceps mishap at birth,* but for almost a hundred years, the Chamberlains made a fortune as 'man-midwifes' by keeping

---

* *Tristram Shandy* was written in 1759, when just about everybody who was anybody had syphilis. One of the classic stigmata of inherited syphilis is a flat nasal bridge, so I will leave the reader to judge if author Laurence Sterne in fact made the first known skeletal pathology joke.

their invention a secret, the potential to profit from their reputation as miracle workers vastly outweighing the potential for general public good. This sounds scandalous, though it is not terribly different from the system of medical patents at work today, and we still see great gaping differences in birth outcomes between the haves and the have-nots. One of the most shocking statistics on birth outcomes isn't the chainsaws, or even the burnings at the stake – it is that maternal mortality is up to *five* times higher in women of colour than their white counterparts in today's America; and four times as high in the UK. Social and economic status remain the top predictors of birth outcomes around the world, which tells you everything you need to know about how far we (haven't) come from the Chamberlains and their secrets.

We know that births have been difficult for a long, long time because we can find evidence of the worst outcomes in the archaeological record. There is a skeleton in the rather charming, if unassuming, town museum of Hitchin in Hertfordshire, UK, that tells a quietly bathetic story. The skeleton is wide-hipped, with a delicate* skull, so we know she was a biological she. The other way we know that the skeleton was biologically female is that she was buried with two infants alongside her – and a final neonate still in her pelvic region. Around 2,000 years ago, this woman died on the cusp of becoming a mother and her children with her. Triplets, of course, are risky, even today. The earliest known case of death in childbirth comes from nearly 8,000 years ago in Siberia's Lake Baikal region in a woman from a hunter-gatherer group who died giving

---

* Delicate-ish. Roman period Englanders were rather more robustly built, as they presumably actually did things other than go to museums all day.

birth to twins. While these are both cases with more babies than human women typically try to birth at once, there are also examples from around the world where even single births have ended badly.

We have more evidence for deaths in childbirth from later periods, like the woman and her foetus who were buried some 6,000 years ago near Henan, China, probably both victims of the woman's overly long pelvis. There are many such cases of foetal remains found either *in situ* in the mother's body, or newborn next to it, and they occur around the world. When a large sample of Chinchorro mummies was examined forensically, for instance, a full 14 per cent of the group of 187 were found to have died of complications of childbirth, mostly related to infection. We only know this because of the preservation of soft tissue, as only a handful of them suffered conditions that left any mark on the skeleton.

For many years, archaeologists argued back and forth about whether human birth really was that difficult in the past, especially before we invented farming and the extra calories that come with it. The Neolithic period, with its sedentary lifestyles and carb-based diets, it was argued, allowed for even more energy to be transferred to babies in the womb. This should by rights lead to bigger babies – and even more difficult births. However, recent estimates of mortality in late Stone Age groups from South Africa show that women were still very much at risk around the time of their first birth. We certainly see a greater number of deaths in childbirth in skeletons after we take up settled lifestyles, but we also see a greater number of skeletons overall. By the Neolithic period, no matter where you are in the world, there are just more people. And this is at the heart of our success as a species – despite all of the obstacles, we have become very, very good at making more of ourselves.

The thing is, birth is difficult and people die. We would not care so much, or have so many weird rules about it, if it wasn't so dangerous and yet such a critical part of our species' survival. But because birth is important, we are still messing around with it. Our species is very keen on taking the important things in life and messing them about with our niftily adaptable evolutionary lever: culture. Sometimes this is helpful, sometimes it is not. In the past, our cultural habits allowed us to create the networks of support human women need to give birth. In the past, but also right now for many women, our cultural habits have had quite a good go at making birth even worse than it already is, in ways that our technological innovations are struggling to adapt to. I mean, birth is still not easy for our species, and yet we can send cars into space for advertising.

If childbirth is so risky, why not give it up? Adapt and evolve and change into a species that doesn't cark it trying to reproduce? Well, that's not part of our investment strategy. We have traded the risks of dying and leaving our existing children without support with the rather less immediate threat of having fewer children overall. Recall the last chapter, where we discovered human mothers are growing quite a lot of their babies *outside* the body. We certainly don't cook them to done – there are no newborns gallivanting around bipedally, following the herd. The type of baby we have, and the cultural and biological adaptations we have made for raising them mean that the hard work actually comes next, bringing up baby. Birth is only the beginning of a lifetime of investment – and that is what makes us human.

## CHAPTER NINE

# Bye, Baby Bunting: Caring for a Child the Old-Fashioned Way

Bye, baby Bunting,
Daddy's gone a-hunting,
Gone to get a rabbit skin
To wrap the baby Bunting in.

W e are not barn spiders, ready to retire from the hard work of childbearing the second our babies emerge into the world. We are mammals, and worse, we are primates. Membership of these somewhat demanding clubs requires that we invest in the physical growth of our children after they are born, to make up for our poor energy-transfer dynamics and outsized baby ambitions. Alongside the physical investment, parents are also obligated to dig deep and invest in teaching their offspring the rules of the jungle, building up social capital in their youngsters so that they have the best chance to navigate their societies and thrive.* This is what mammal, primate and human parents all have in common: when it comes to our kids, we have to *care*, or the whole thing falls down. Which is why it is particularly shocking to see that when history looks at children in the past, there is a sort of general assumption that humans don't – or didn't – really particularly care for their children.

This is patently nonsense. Alongside the very earliest written words imploring gods for protection and support

---

* Oh, the social capital you will lose if you become a parent.

we find all the ephemera of anxious parenting – the tools for feeding, bathing, swaddling and playing with a baby. Three-thousand-year-old amulets adorned with the monstrous figure of a woman with a lion's head and bird's talons plead with the pantheon of gods to protect pregnant women and newborns from the ravenous goddess Lamashtu. We can read that, like parents everywhere, Aztec parents in the past called their babies pet names, albeit culturally specific ones such as 'precious necklace', 'precious feather', 'precious greenstone', 'precious bracelet' and 'precious turquoise'. Despite the strange, morbid assumption that seems to have prevailed in a certain kind of history over the past few hundred years, these people *cared* if their individual children suffered and died.[*] Without the benefit of having read Hobbes or Malthus, they were not necessarily aware that they should accept childhood mortality as a statistical quirk of the population density need to underpin urban economic function. The parents who set up the headstone for their eight-month-old daughter in the city of Mainz in the ancient Roman period were obviously ignorant of how they should behave when they carved their lamentation: "Oh, had you never been born, when you were to become so loved, and yet it was determined at your birth that you would shortly be taken from us, much to your parents' pain … The rose bloomed and soon wilted."[†]

How foolish they must have felt, not knowing that by the time learned, and, one suspects, childless scientists began to discuss the history of childhood in the nineteenth

---

[*] This seems to be an offshoot of the nasty tendency in a certain school of history to view the past as a different country, and to view different countries as full of people who are not, actually, really people.

[†] See reference in Carroll, 2018.

century it would be clear that Romans did not mourn their children if they died under the age of a year, because august authorities like Plutarch and Pliny said so and they were of course perfectly unbiased chroniclers of the human condition, and not at all grumpy old men with an obsession for arbitrarily rigid and unsentimental social rules that could never actually be enforced in a real society.

It is really very bizarre that the idea of childhood in the past has come down to us stripped of emotion, and we can learn that there were no set funerary standards for Roman babies, for instance, and read into this that Romans did not care if their babies died. Similar intimations have been made for the children who died throughout the ancient world; from Spartans exposing 'defective' children or Carthaginians massacring babies.* Bioarchaeologists Helen Gilmore and Siân Halcrow have compared, I think rightly, the way we look at the past with the way early anthropologists looked to the 'exotic' and 'primitive' cultures of the world around them; even today there is a tendency for archaeologists to rewrite the past as something other, something foreign, and full of darkness and mystery that we cannot imagine in our world today. It says very little for our empathetic capacity, let alone our ability for self-reflection, when we attribute inhumanity to the unknown past, and harks not just a little of the nasty taint of colonial sentiment that *our* sensitivities are heightened and refined, while *theirs* were nasty and brutish.

Happily, some of that self-reflection has started to pay off within the discipline of digging – and studying – the people of the past, and we are able to revise some of our harshest judgements on the people of the past.

---

* And that one comes loaded with modern ethno-nationalist issues.

A prime example is the battle for the majestic north African naval power of Carthage: two rival teams of archaeologists and anthropologists locked in a now decade-long battle to decide the truth of fabulous accounts of child sacrifice. Plutarch* is only one of the writers to accuse the Carthaginians of sacrificially feeding their babies into giant metal statues with fires burning in their bellies. Admittedly, it all started off looking very grim for the Carthaginians. In the 1920s, thousands of cremated remains – all of young children and infants – were discovered in the Tophet cemetery just outside the ancient city of Carthage. The fact that these young children had all been cremated, while none of the other burials in Carthage had, gave weight to the fed-to-the-god-Moloch hypothesis.† However, bioarchaeologists, professional analysers of the dead, have additional ways to investigate the truth of such rumours; through the evidence of the skeletons themselves.

Cremation, contrary to popular belief, does not destroy the whole of the skeleton unless incredibly high temperatures are reached; even modern cremation practices with industrial heat sources leave some elements of the skeleton behind. At Carthage, the question is whether these were the babies of citizens, sacrificed wailing and screaming to the gods as the historians insist: are the remains in the urns children, and did these children burn to death? Extreme heat does destroy quite a lot of evidence that would normally be used to determine both cause of death and the age of the deceased, both of which would give us some clarity as to who was being buried in the Tophet – and why. Luckily for science, however, these cremation burials never achieved the heat needed to actually fully cremate a

---

* Yes, him again, miserable sod.
† And Plutarch's dim view of humanity (*On Superstition*, Chapter 13).

Figure 9.1. *Illustration of 'Molech' in* Bible Pictures and What They Teach Us *by Charles Foster, 1897.*

skeleton – several of the sturdier elements of the body, like bones and teeth, are preserved.

Given that modern science knows the approximate size of a baby's bones, it should be a simple thing to measure the length of the surviving bones, and plot them against age and growth charts to establish whether these are indeed newborn children. The first set of results to emerge from such analyses suggested that indeed, these were babies in the urns, but they were very young – a little less than a third of them *too* young to have survived outside the womb. This led to the supposition that at least some of the burials in the Tophet must have been infants who died of other causes, and that the cemetery may have been more of a special place reserved for infant burial rather than sacrificial burial particularly. The pro-child-sacrifice researchers countered that it was impossible to judge age from bones that had been quite probably warped by cremation/statue fire and, importantly, shrunk by heat. Cue rival headlines in the sorts of newspapers that report on 2,000-year-old rumours of child sacrifice and a seemingly unanswerable question.

Unanswerable, that is, until you look straight into the jaws of the issue: into the teeth, their high mineral content allowing them to survive the heat and step into the archaeological record to tell a story that the rest of the body can no longer deliver. One thing we know, and I do mean *we* in this case, because this is something I have been researching for years with my esteemed colleagues at the Natural History Museum, is that the one trauma you can reliably count on to scar teeth is birth. Being born is sufficiently traumatic that a baby's body sort of stop-starts normal growth, and that stop-start shows up as a *neonatal line*, which is a scar through the inside of your teeth where all the little cells had their moment of existential panic. It exists in all of your kiddie teeth and even in ones you have as an adult – the first big chunky adult chewing tooth that comes in was actually forming before birth, so you can see the scar there too. This scar allows bioarchaeologists and forensic scientists to establish whether a baby was born alive, and in 2017 the not-all-sacrifice team clearly demonstrated that half of a subsample of the infants buried at the Tophet had no neonatal line in their teeth. So, while we cannot say for certain that Carthaginians didn't delight in throwing their children

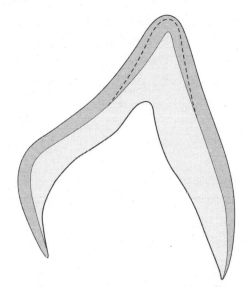

Figure 9.2. *A histological section through the central axis of a baby tooth –
in this case a canine – will reveal a mark that formed at birth (dashed line).*

into the fiery maw of Moloch or the goddess Tanit – and
certainly there is overwhelming, if biased, historical
testimony that they did – we can say that of the thousands
of infants cremated and buried outside Carthage, many
may have been the victims of a much crueller force:
natural mortality. But it certainly doesn't mean their
parents didn't care for them.

The issue of how much people actually care for their
children – and whether they might care for older children
more than younger – is one of considerable antiquity.
Plutarch informs us (as much as anyone could 600 years
after the fact) that the legend of Spartans leaving any baby
not deemed to be perfectly formed to die of exposure is in
fact based on actual practice.* Enough historical sources

---

* Him again. Not given to cheer and goodwill towards all men,
Plutarch.

agree with him that we can understand that exposure was a real threat to the newborn in ancient Greece, but estimates that 10–20 per cent of all babies were left to perish in the days after birth are way over the top – that is simply too many babies to lose in a normally demographically functioning society. There are many reasons why babies would have been unable to be raised in ancient Greek society; many of the textual accounts seem to imply that exposure was the fate of babies who were already not thriving, while other reasons include illegitimacy and possibly sex – in a system where daughters need dowries, there may be pressures towards selective infanticide of expensive female children.

That isn't to say the trauma of losing or giving up a newborn wouldn't have been precisely the same as it is today. Rather, that there were different social rules for such young babies and either nature or circumstance could drive parents to extreme action. Of course, for the Greeks, exposure wasn't the same thing as infanticide and that's the bit that people don't pick up on. Later historians have argued that many of the children 'abandoned' by their parents were indeed adopted into new families, and that such an abandonment may have bettered their chances at survival. Alongside the tradition of exposure we have a tradition of rescue and adoption – if Greek moralisers enjoyed chastising their fellows for having children they could not raise, then Greek dramatists at least delighted in saving those doomed children for some more poetic fate.

Sometimes, care is a little harder to understand. There are practices of care in the past that would pretty quickly get social services called on you if you tried today. Recall from Chapter 7 that parents of yesteryear so valued their children that they spent considerable time and effort squishing their babies' heads into weird shapes. The soft, unfused bones of a child's skull might be wrapped with a series of fabric bands, forcing growth either higher up or towards the back of the skull and away from a stylishly narrow-waisted middle; flat boards might be tied sandwich-frame-style around the infant's head to pinch the skull into the high, narrow shape that television programmes about ancient aliens are so fond of. Modification of baby skulls into socially pleasing shapes has been practised by cultures on all the inhabited continents, from as far back as 13,000 years ago in Australia at Kow Swamp on Yorta Yorta land, 11,000 years ago in north-eastern China and about 10,000 years ago in Mesoamerica.*

Intentional cranial deformation may seem to us a rather peculiar way of combining investment in both the child's physical and social well-being, but it is not so far off customs like ear-piercing and circumcision that are still practised today. Researchers have argued that intentional cranial deformation can be a marker of social status, which makes sense if you think of it as a practice that must be deliberately pursued for years in order to achieve the ultimate effect – it is just another type of investment parents are willing to make. The time it takes to shape their little heads into a community-acceptable standard is yet another example of how humans can physically embody their investment in their children.

Given that when we think about children in the past, we tend to imagine a harsher, Hobbesian world, where necessity stripped emotion and care from the business of childhood, it is perhaps surprising to realise how little

---

* www.biorxiv.org/content/10.1101/530907v1.full.pdf

evidence we see of actual maltreatment and physical
trauma. Mary Lewis is an expert on the bioarchaeology of
children and has tackled the difficult subject of child abuse
in the past, but even her research has only uncovered a bare
handful of cases, less than could be counted on the fingers
of two hands, of the traumatic fractures such an uncaring
world would invoke. Abuse, of course, is a difficult category
to pin down, given that attitudes in many societies about
what constitutes abuse have changed in only a matter of
generations.* But the vast amount of evidence we see for
childhood in the ancient world is that of parents trying to
do the best for their children and preparing them for
adulthood. Where once we saw high infant mortality in
the past as potentially desensitising – while simultaneously
assuming every cemetery of infants must be due to murder,
rather than cultural rules for burial – we now have an
entire subfield dedicated to understanding the great lengths
that human beings go to in order to keep their loved ones
alive: an archaeology of care. We have a vast array of
material culture – those *things* that so clutter up our
archaeological record – that speaks of the love and
investment parents have poured into their children.

Which is not to say it's easy. This leads us to the trap
every human parent encounters: the worry that you're
doing it wrong. Confronted with an obviously glitching
iteration of our species, whether it is crying, wailing or
simply giving you a look, it is no wonder that we have
come up with any number of proscriptions on how to care
for our babies. How do we do it and – this is the question
that launches a thousand books – is there a right way and

---

* As an example, ask someone born around the middle of the
twentieth century their immediate associations with the words 'get
the' and 'belt', 'strap', 'switch', 'hiding', *etc.*

evolutionary way, to do it? Is there in fact an Ur-childhood, a palaeo-parenting scenario that is optimally adaptive, perfectly responsive to our evolutionary needs? The answer to this is a rather resounding 'no', unless of course the species has completely died out by the time I finish this manuscript,* and condensing a few hundred thousand years of existence into one typical experience suddenly seems expedient. Otherwise, of course not. We are a constantly adapting living species and, as a going concern, we are actively changing the way we interact with our environments through genes, behaviours and humanity's great mediator, culture.

Care and feeding are the monolithic juggernauts of the evolutionary parenting armada, but – spoiler alert – the way humans do them are culturally determined. The way we care for our babies now is dramatically different from what was probably the experience of the vast majority of our ancestors. We sleep separately, we drink artificial milk and we worry that neither innovation is beneficial. But Darwin's evolution is a simple concept and one we have managed to turn into a strange sort of secular godhead, asking at the altar of 'survival of the fittest' if our lifestyles are evolutionarily adaptive. And there is no one more willing to look for guidance than an anxious parent.

Are our baby-care strategies adaptive? Well, what are you trying to adapt to? The number of human beings who live by hunting game, collecting shellfish or foraging for their food is small and getting smaller. And groups that we

---

* Large parts of this book were written under lockdown conditions prompted by the spread of the novel coronavirus that causes Covid-19, so honestly, I'm more ambivalent on the end-of-days issue than previously.

regularly turn to for reference to some perceived 'evolutionary' lifestyle, who we call 'hunter-gatherers', are no
more set in their ways than you are. They exist in a world
of environmental devastation, encroaching development
and the inexorable march of capitalist systems across even
the most hostile of landscapes. The various people living
non-agrarian lives mentioned throughout this book do
not have uniform experiences any more than any human
anywhere does – and they do not live in some pristine
unchanging world. Many are involved in wage-labour
economies or have moved to permanent settlements.
Almost all forager groups, from Brazil to Tanzania, have
been pushed into increasingly marginal environments by
the pressure of agrarian states, some within living memory
and some over the course of the past several thousand
years. The foraging groups we talk about when we discuss
'palaeo' lifestyles are no more adapted to the Stone Age
than we are. They are modern people adapting to very
specific modern circumstances – and this is a good thing
to remember when someone is trying to sell you a palaeo
snack bar.

This infuriating insistence on a single, stable human
culture over hundreds of thousands of years is forgivable,
perhaps, in the competitive world of protein-bar marketing,
but it has no place in deciding how you raise your children.
Another word for achieving an unchanging, stable state is
death; human societies no more achieve a viable stability
than individual humans achieve immortality. We change,
we cause change, we adapt. We have needed to adjust to
new climates as we spread out into the world, across the
radically different ecotones of our native Africa to some of
the most extreme environments on earth. We have had to
cross land bridges, mountain ranges and entire seas to
establish ourselves on new continents. The human

achievement in populating Australia some 50,000 years ago is hardly going to have operated under the same cultural conditions that saw modern Europeans' ancestors chat up Neanderthals in the Middle East.*

So, if human culture cannot be assumed to have remained static for all that time and through the myriad changes we have made, then can we really say what an 'evolutionarily adapted' (no) or (never say this) 'palaeo' childhood would have been? Because a great deal of people seem to believe that they can. It is, on the one hand, incredibly reassuring that the lexicon of evolutionary thought has so permeated our society that people actively seek explanations for our modern behaviours and conditions within the framework of generations of adaptation. What is less reassuring is that almost all of the explanations that percolate into public consciousness have all the evolutionary adaptability of a dead parrot. Evolution is presented as something that happened in the past and usually a specific part of the past. Our awareness of it is highly selective: we mostly talk about how our species would have behaved in the savannah environments we think we started occupying around 2 million years ago, but bypass the much longer time our ancestors spent in forests – anything with a roof on it doesn't even get a look in.

There has been an awfully long part of our time on this earth that has been occupied living in ways we (mostly) no longer live today. Ninety-nine-point-something per cent

---

* Though the first Australians may have encountered *Homo floresiensis*, the hobbit, on their way out. Which is rather awkward for everyone as *floresiensis*, much like Neanderthals, disappears from the archaeological record after contact with modern humans. This means we have been literally ghosted by more than one species, though, knowing us, it's hardly that surprising.

of our time on earth has been spent living in small, free-ranging groups; we didn't even make it as far as the Americas until around 20,000 years ago. It makes intuitive sense to us that such a simple existence would be an unchanging one, and therefore the perfect evolutionary sandbox in which to build an 'ideal' human lifestyle and 'ideal' child-rearing practices. This is the reasoning behind every palaeo-branded item any huckster has ever tried to sell you; the Edenic idea that we once had it all sorted and, merely by eschewing bagels, modern humans can somehow recapture a glorious prelapsarian state of equilibrium, health and glossy hair. The pernicious thing, however, is that palaeo-ideal phenomenon is not restricted to dietary supplements and bizarre workout routines, however. The idea of a 'one true strategy' for living also finds its insidious way into the advice we give on child rearing. The concept of an evolutionarily superior way to raise kids places a crushing weight on the shoulders of parents to do things 'the right way'. When such concepts come in 'scientific' wrapping, it is even harder for the non-specialist to unpick what is a long-term trend in human evolution – and what is merely so much monkey shit.

Sarah Hrdy, the primatologist and anthropologist encountered in Chapter 4, has pointed out that much of what we think of in terms of 'correct' ways to parent is in fact the product of short-lived cultural trends[*] and the introduction of evolutionary perspectives to parenting is not necessarily straightforward. Many parents, and innocent bystanders, will have heard of the 'attachment

---

[*] The cult-like adherence of parents to the work of Dr Spock (the paediatrician, not the Vulcan) would be one such example, shifting attitudes in a matter of two decades from the hard-nosed negotiating parent of the 1950s to the compassionate cuddler of the 1970s.

theory' of John Bowlby and Mary Ainsworth – that babies are hardwired to cling as hard as possible to their mothers and this is an evolutionarily adaptive trait, offering security and protection. 'Attachment parenting' is now a very strong trend in the developed world, based around the work of William and Martha Sears, and carries highly proscriptive instructions for the caregiver based around instant contact after birth, continuous physical contact throughout infancy, including when sleeping, and heightened awareness and responsiveness to the baby. Many people will be familiar with this trend through the phenomenon of 'baby-wearing', or keeping an infant strapped to the body of the parent. Baby wearing is lovely and excellent for freeing hands, but taken to extremes it means a new parent is now expected to be able to rapidly manipulate a large and expensive band of cloth into a variety of knots and ties that will hold a live animal of changing weight and agreeability.[*]

This is a strange reversal of the odd evolutionary trajectory that eventually rid of us the fur primate infants need to cling to – though, perhaps not so much a reversal as a new direction, given we haven't quite yet reverted to stashing our babies in tree trunks like our presumed ancestor the tree-shrew-like-thing. This all-or-nothing attachment parenting style, which in its most extreme forms demands total subjugation of the mother to the emotional needs of the infant, may not actually reflect reality of how human attachment works. The types of attachments Bowlby identified as being important to infant development through his early research on primates only

---

[*] Buy a baby carrier to strap to your chest? That's very modern and modern is just cheating, isn't it? Though, from experience, the baby is unlikely to notice.

came from the mother, because he studied species in which the mother was the only caregiver. Humans, with their endless resources of alloparents – grandparents, relatives, siblings, hired help* – may be able to be far more flexible in creating attachments than baboons.

However, this kind of ambiguity even in the most accepted of parenting concepts has not stopped the onslaught of concern that there is, somewhere, a more adaptive way to parent. One brand of palaeo-parenting – and it is a brand, they are selling you something – actually called 'evolutionary parenting', 'focuses on the idea that any time we deviate from a known biological norm, we should have good reason and try to mimic biological processes as much as possible'. This is excellent advice, if the issue you are dealing with is, say, heart function. Hearts have a very specific range of biological states they can be in before things go very, very wrong and we should absolutely try to keep our hearts (and those of our children) processing blood within the very narrow tolerances a human body can take. If we can't do that, we should employ devices that help us mimic the normal operation of a heart, like pacemakers.

This kind of statement is absolutely meaningless, however, when it comes to something like whether you should keep your baby strapped to your chest at all times – a fairly frequent subject of debate in the endless cycle of competitive acrimony known, infuriatingly, as the 'mummy wars'.† The human baby is capable of surviving strapped to your chest

---

* For instance, my parents opted to leave their children with a woman whose only real fault, realised sometime after the fact, was that she was the wife of the biggest cocaine dealer in our city.
† The one instance of territorial aggression we are happy to attribute to women – and it involves sniping about natural fabrics.

quite a lot of the time, some of the time or none of the time. A baby's biological norm is a set of human contact, temperature and pressure requirements that, unless you find yourself suddenly dropped into the atmosphere of Venus, can be met in an absolutely dizzying variety of ways. Babies who have gone on to lead full and happy adult lives have been positioned strapped to chests, strapped to boards, hung on walls, in their parents' arms, in slings or even – and this is a personal favourite – strapped to a board and *then* hung on a wall.

The idea of a biological norm is a pernicious one, however. There are aspects of infant care that deal directly with meeting the biological needs of a growing baby; these occasion the most concern and, unsurprisingly, the most vitriol among those who envision that there must be a *correct* way to meet these biological needs. It would save the internet an immense amount of data if we were to point out that meeting biological needs isn't the same thing as the baby having a specific way that biological need has to be met, but online cultural disagreements rarely allow for that kind of nuance. Instead, there is an endless anecdotal stream of advice on how to care for your baby and a fair amount of performative martyrdom. And there is no one more willing to take advice, any advice, than a sleep-deprived parent.

Some 4,000 years ago, a scribe working just outside the ancient city of Nippur, which we would find today in modern southern Iraq, set down in neat, angular lines on clay the words of an incantation. Buried and lost for thousands of years alongside other such magical texts, the clay tablet carrying the incantation eventually came to light in the twentieth century, when it was painstakingly translated from Akkadian, though the words are so prettily arranged that it is likely they come from a far

older folk tradition, sung as a lullaby. The magic it begged
for was something that will be familiar to any parent of a
new baby:

> Little one, who dwelt in the house of darkness;
> Well, you are outside now, have seen the light of the sun.
> Why are you crying, why are you yelling?
> Why didn't you cry in there?
> You have roused the god of the house, the kusarikkum has
>    woken up:
> 'Who roused me? Who startled me?'
> 'The little one has roused you, the little one has startled you!'
> As onto drinkers of wine, as onto tipplers, may sleep fall on
>    him![*][†]

Sleep, we can agree, is a biological need. Babies need sleep
to thrive. How they sleep, however, turns out to be neither
here nor there. This is a bit of shock, given the vast quantity
of column inches given over to warning new parents that
their babies will grow up to become sociopathic monsters
if they sleep apart from a parent, or that they will grow up
to be pathologically dependent if they are constantly
coddled. If we were one of our tree-shrew-like ancestors,
perhaps we wouldn't have these kinds of issues. Our babies
would mostly sleep in their very own nest, apart from
their mother, who would be a single parent mostly
concerned with the safe parking of her infant while it
grew big enough to forage on its own. This single-parent
parking evident in tree shrews might well be the base or
ancestral condition that all of our subsequent sleeping
behaviours have derived from, though of course you'd

---

[*] Proof, as if any was needed, that praying for a baby to sleep (like a
drunk, in this case) is of considerable antiquity.
[†] Trans. by Faber, 1990.

have to ask the tree-shrew ancestor itself – and it hasn't
been around for several million years.

Most primates don't park and they don't all sleep alone.
The advantages of living in a group extend to sleeping in
one – there is a better chance of predator awareness and
defence. Baboons, for instance, sleep in a huddle at ground
level, relying on their numbers to keep them safe from
marauding predators. While most primates opt for the
arboreal sleeping option, there are disadvantages to trying
to accommodate your entire group in a tree. Solitary
living has given way in most of our clade to group living,
and you are just not going to fit that many monkeys into
a hole in a tree. Primates who are arboreal group sleepers
by necessity have to accommodate themselves across
branches, which are neither good for protecting against
aerial predators or, according to field reports of falling
monkeys, sleeping in. Another challenge comes with size.
Something with the heft of a gorilla is not going to
manage to nap in anything but the sturdiest of trees. Great
apes, by definition on the large side, require additional
structural support: gorillas, chimpanzees, bonobos and
orangutans all construct sleeping nests or platforms for
themselves.*

In almost all cases, however, whether it is in a hole in a
tree or a carefully constructed nest of twigs and leaves,
the primate baby sleeps with – or on – its mother.
Depending on the size of the bed and the animal's
propensity for pair bonding, group living or general
sociability, it might be joined by its father, siblings, aunties

---

* Gibbons, the smallest great ape, do not make sleeping platforms.
They are, however, very tree-focused, and seem to manage branch
sleeping without major incident. My toddler, by contrast, will roll off
any surface given half a chance.

or uncles, or any combination of group members. One of the reasons for this crowd is the adaptation, in many of the primates that are on our end of the family tree, for life in the daytime.* Shorter bouts of sleeping, in larger protective groups, seem to be a consequence of a transition from nocturnal tree rat to grounded and gregarious. The sole case where baby sleeps apart from its mother in the great apes occurs in a handful of cultures of anatomically modern humans.

But what does it mean to sleep apart? In primates, co-sleeping is very physically constrained – in a tree hole or nest there are not a lot of options. In humans, however, there are a variety of ways of viewing 'co-sleeping'. The big cross-cultural survey of Barry and Paxton compiled data on mostly boy children from 186 global cultural areas on all manner of things, from who carries the child to where and when the child learns to walk. They report that, globally, about 79 per cent of cultures sleep with baby in the same room as at least one parent, while 44 per cent actually sleep in the same bed. Others have argued that an even closer form of co-sleeping, 'breast-sleeping', should be considered the human norm, where the child sleeps in direct contact with the mother in a position that facilitates breastfeeding. They suggest that what Western or European medicine has established as a biological norm – sleeping apart from an infant – is in fact only a very recent invention and that it may be so removed from normal that it is actually maladaptive.

The prevalence of co-sleeping among modern foraging groups has led to the suggestion that our human ancestors who also made their living without settling down in one

---

* While it may not be obvious to owner-operators of small infants, human babies are diurnal animals.

permanent location would have reached the same solutions to the problem of where to sleep. In general, people from foraging groups tend to sleep more to a bed, and with their kids until children are much older, than in groups that utilise different subsistence strategies. Indeed, in the campsites of modern foraging groups like the Hadza of Tanzania, the average number of people in a bed is 2.4. You may not imagine that as either practical or conducive to a good night's sleep, but of the small group of families studied, only one father in the whole camp was found to have given up on co-sleeping with his family.*

The clearest argument for assuming that co-sleeping would have been the norm for our ancestors is that the separate physical space of a nursery, where infants and children would sleep without their parents, was not really possible until you had private dwellings with multiple rooms. It is very difficult to imagine a new parent in the Upper Palaeolithic patiently collecting mammoth tusks to build a nursery addition to their tusk-shelter when a baby comes along. It is probably safe to assume that prior to the invention of homes with multiple rooms we were all habitual co-sleepers. Whether the invention of rooms was truly pathological, for sleep or for anything else, is a subject for another time. For now, it is safe to say that the evidence from modern ethnography and evolutionary anthropology all suggest that, at one point, human babies slept with their mothers — and possibly their fathers, siblings and other

---

* And he said it was the heat, not the kids. The paper does not indicate whether this was said within earshot of his family.

invited guests too.* The important thing is that whether
you particularly want to sleep with your baby lovingly
kicking you in the kidney all night, or whether you have
the wherewithal to build a mammoth tusk extension to
ensure you have some peace and quiet, the way we humans
adapt is through our culture, which basically means it's up
to you.†

---

* Important note: there are very good reasons that Western/European
medicine warns against co-sleeping, which have to do with risks to
the baby that are simply not the same as they were in the Upper
Palaeolithic. Take the best medical advice you can on these things – as
has been pointed out repeatedly, I am not a 'real' doctor.
† Or rather, it's up to your baby. Seriously just get some sleep,
however you can.

## CHAPTER TEN

# Old Mother Hubbard's Cupboard: the Magic of Milk

> Old Mother Hubbard went to the cupboard,
> To fetch her poor dog a bone.
> But when she got there, the cupboard was bare,
> And so the poor dog had none.

There is care and then there is feeding. Human babies require a very specific kind of feeding, one dictated by our evolutionary history not just back through the tree full of primates, but way, way back, to the beginning of mammal life and then further, further back to the invention of post-birth care. We are investors, nesters; not for us the eating of mother's corpse when we could get so much more out of her alive. Humans take the investment we have made in our offspring and we double, maybe triple down after birth. Any number of animals follow this strategy of course – there are birds dropping worms into baby beaks and lion cubs being helped to pounce. Care is an investment many species make. When the investment is in the physical well-being of the child, we are happy to see it embodied across the animal kingdom as its own kind of capital. A fat young robin is the result of some slightly less fat parent robins, who have given up something of themselves to put into their offspring. Mammals are no less keen than robins; what's more, we even have our own adaptive strategy for investing in our offspring's growth after baby is born. This is the miracle of milk, and milk, it turns out, has a lot to say about our evolutionary history.

The mammalian answer to 'how do I grow something as big and impressive as myself' is a fairly incredible secretion from specialised glands that manages to feed and water our offspring in an incredibly responsive and highly adapted way. One of my absolute favourite milk researchers,[*] Katie Hinde, whose work informs a lot of this chapter, refers to milk as a 'biofluid', which reflects that what we think of as something static and uninteresting hasn't been handcrafted by evolution for millions of years just to sit in your fridge all unchanging. Milk *does* stuff. All mammals go through a stage after birth of exclusively consuming their mother's milk. However, all milk is not created equal;[†] some milks are hugely fatty, some are sugary, some are watery, some have this protein, some have that. Some are designed to be a short-term solution to a big growth spurt, and some seem designed to go on far longer than seems practical. Different adaptive strategies have led to different kinds of investment in milk across mammal species, including decisions somewhere down the line in human evolution whose consequences we are still living with today.

Milk is designed to do one thing – get a baby from womb to functionality, and there are only so many ways to push nutrients out of one animal and into another. We humans branch out and drink the milk of other species – cow, buffalo, yak, horse, donkey, goat, sheep, camel and even reindeer.[‡] However each species has a

---

[*] Life is improved massively by having favourite researchers.

[†] Soy milk. Ugh.

[‡] If it can be milked, humans have probably milked it. Animal rights campaigner and former Beatle-wife Heather Mills once suggested that rat milk was likely the only ecological solution to the serious ethical and environmental issues raised by human consumption of dairy. It is unclear how the rats would feel about this.

specific milk recipe that has been honed down the generations to provide optimal nutrition in a way that fits in with a hectic mammal lifestyle. A different balance of fat, protein, water, vitamins, all prepared for delivery in specific, optimised ways. Think of it this way: if you are a mama hooded seal and have exactly four days to nurse your pup out on the Arctic ice before something eats you, you bet you're going to have milk that tops 60 per cent fat because you need to grow that baby *fast*. Milk fattiness is linked to the overall density of the energy the mother is trying to get into that baby. If your babies are instead going to be nursing while you're in hibernation shutdown mode like a polar bear, you might keep back some of those critical nutrients to see your own giant furry carcass through the winter.

Milk is designed to meet the very specific needs of a specific infant at the exact time it is produced, which is perhaps not what you want to think about as you stir commercial cow's milk into your morning coffee.[*] Every species has its own special formula: those hooded seals are making incredibly fatty milk because of the success of those polar bears in making their milk last long enough to grow more terrifying, seal-eating polar bears. According to Katie Hinde, humans produce milk that is about 4 per cent fat, 1.3 per cent protein, 7.2 per cent carby-sugary lactose and 90 per cent water. You know who else produces milk like that? Zebras. As she rightly points out, we are not zebras. We don't gambol across plains wearing stripes, eating grasses and, according to the science writer Jared Diamond,

---

[*] But probably should. Large-scale dairy production is not ... nice. For cows, or the environment.

being murderous.* We occupy an utterly different ecological niche, have completely different schedules for the important events in our life histories, and generally are just not very closely related. Yet our milk is the same, because the milk you make reflects the way you plan on distributing it.

The contents of mother's milk adapt to circumstance and to need, both on the big wide species scale and on the individual baby scale.† As a rule, wandering mothers produce richer, less watery milk. This makes sense: if mothers have to travel away from baby to find food they aren't able to nurse as frequently and so had better deliver as much nutrition as possible when they do. On the flip side, mothers with babies in close range might be able to offer far more diluted stuff, because they are pretty much available for feeding whenever. For primate babies there are two basic options: it's park or ride.‡ Some primates – mostly those down far away from our particular anthropoid branch, like our squirrelly buddies the galagos, carry their babies around in their mouths when they are very little, like particularly adorable monkey-kittens. These mouth-carrying species of

---

* Ok, three out of four. Though I'm not sure I believe Diamond about the murder-zebra thing.

† Human babies, for instance, can prompt their mothers to produce milk pretty much on demand by simply *looking hungry*. The mother can then be carefully trained to produce on a schedule amenable to the consumer with a series of carefully timed cries, and woe betide the mother who does not realise when meal times change because being too full of milk is painful and unsustainable. It also adds unnecessarily to the laundry pile.

‡ Hats off to the primatologist who used this as a paper title on primate infant care. Not quite up to the levels of the five Swedish scientists who dedicated the latter part of their research careers to Bob Dylan songs, however. Who doesn't want to read 'Nitric oxide and inflammation: the answer is Blowing in the Wind' or 'Tangled Up in Blue: molecular cardiology in the postmolecular era'.

primate tend to park their kids for long periods, which seems sensible given the difficulty of leaping through their arboreal — or at least quite tree-saturated — habitats with a mouthful of baby. Galago babies are then left in said trees in the same way Bambi was left in that forest glade and, as a result, mama is able to range quickly and efficiently, if occasionally tragically, for food. In fact, threats to parked babies — not necessarily from the malign imaginations of Disney animators, but certainly from predators like barn owls and other birds of prey — may have spurred the development of the alternative: the clinging, or 'riding', infant. For parked babies, however, milk needs to deliver the goods as fast as possible, because mama has to be on the move and doesn't want to attract predators. Parked baby milk is high fat, low water; different species of baby-parking galagos and lorises, for instance, have milk that is between 7–13 per cent fat (compared with our four).

We (and zebras) are accustomed to being in close contact with our mothers more often than the latchkey lorises and it shows in our milk: we have much more dilute, watery milk, which can be given at leisure because the baby is always there. This is easily accomplished for zebras, which are born able to walk after mama just fine, but it gets a little complicated in primates. As we saw in previous chapters, many primate babies are born undercooked and unprepared. They do not walk at birth. Instead, quite a few of them have learned to cling. Clinging to mama as a lifestyle choice is not actually very common in mammals — it's only really popular in primates, (o)possums,[*] anteaters and, despite what one would think were the obvious downsides to

---

[*] The masked North American bandits, not the Australian charmers. Despite science, I refuse to start pronouncing an 'o' that I didn't realise was there for decades.

attaching a baby to a flying machine, bats. For primates that are a little quicker on the motor skills uptake, clinging to mama – or dad – is a viable option for making sure that food and protection are never far away. All of this clinging may even have been the catalyst for developing the clever little primate hands we are so proud of today. It certainly has a knock-on effect on the type of milk we produce.

The arboreal orangutan mother carries her baby clinging to her chest, sort of over her hip, some 85 per cent of the time until it can be safely trusted to manage the high-wire act that is orangutan locomotion – after which she acts more as an ape-y bridge between particularly worrying tree gaps. Chimpanzee mothers, who are far more down to earth, only spend about 12 per cent of their time with a baby clinging to their front in those early months, but will also use themselves as structural supports while the baby explores moving through the trees and, critically, not range very far from their offspring. Both species produce milk that is about 2 per cent fat. The first mode of carrying in knuckle-walking gorillas is an amusing tripedal lope with the baby's chest cradled to theirs, and then later with baby riding along piggyback on their shoulders or back. This is aided by the precise engineering of both infant grip and the mechanical properties of ape body hair – unsurprisingly, bald-bodied humans are remarkably poor at piggyback rides,* and at least one theory (though not a widely subscribed one) has it that one of the reasons behind our rather outré insistence on walking upright might have been the requirements of carrying an infant who has nothing left to cling to. Gorillas, with their weird one-arm carry, however, produce less fatty milk than other primates;

---

* They are also liable to become bald-headed humans should they test an infant's ability to grip and cling to head hair.

around 1.5–2 per cent. The champion for light milk, however, is the bonobo – whose clinging babies won't go more than a few metres from mama for at least a year – at around 1 per cent, theirs is clearly a milk designed for close and constant delivery.

Milk directly taxes the animal that has to produce it. Each baby essentially strip mines its mother in a quest to extract the base materials it needs: carbohydrates like the all-important lactose in breast milk to fuel growth, fat to do what the carbs don't, vitamin D, calcium and phosphates to grow bones and teeth, protein for building complicated tissues and enzymes, and all the other stuff that make up an animal. The component parts of milk – proteins, fat, vitamins, *etc.* – have to come from somewhere and that somewhere is pretty limited to the person doing the milk production. There is a word for the self-cannibalisation that mammal mothers perform to produce milk: catabalisation – and mothers catabalise the heck out of their own tissues, whether it's fat, bone or worse.

Not only does the baby extract energy and vitamins through milk, but milk also transports microbiota, immuno-globulins and hormones, and feeds the growing gut flora in baby intestines. Milk is truly a biologically active substance, not just for growing a baby, but populating and feeding the creatures that live in its intestines, sending hormonal signals, and even blocking and neutralising harmful pathogens. A small but critical percentage of human milk, for instance, has evolved to consist of funny chains of sugars – oligosaccharides – which humans don't digest, but our gut bacteria do; with milk we are literally feeding not only our babies, but our babies' bacteria too.

Milk is yet another way of investing in offspring – and of course, investment strategies differ. Lots of nutrients can be passed from what mama eats, basically allowing her wonderful leeway in consuming more in order to provision baby. I say wonderful, but the 300–700 calories that women burn while breastfeeding are easily accommodated by changes in metabolism through tweaking diet and activity level. This means even if a woman no longer needs the resources she accrued upon her person during pregnancy, it is still going to take some doing to shift them. While some women happily report breastfeeding as a miracle diet, others, like actual sports phenomenon Serena Williams, do not lose pregnancy weight easily, even on a restrictive diet like the sugar-free vegan one she adopted – a diet, by the way, she only had to take up *because* she was breastfeeding. Milk is such an efficient transport system that it can carry with it parts of the adult human diet (animal proteins, wine, espresso) that can be difficult for babies to deal with. It has become very common in well-heeled countries to deal with potential causes of upset by cutting them out altogether, so breastfeeding mothers facing babies with gastrointestinal issues will often cut out some combination of dairy, wheat, meat, sugars, vegetables and fruits of different stripes out of their diet, which Williams dutifully did for her baby.*

Milk by and large, however, does the job of transferring energy from one entity to another with incredible finesse. Fats and sugars from the mother's own reserves are deployed for the all-consuming effort of growing a new mammal baby, even as the mother's bones are being ground up and fed into the baby through the magic of milk production. A Pizza Rat mother, for instance, liberates something like 44 per cent of the calcium in her bones to feed her pups; cats can lose nearly a third of the calcium in their skeleton (if

---

* That sucks.

they don't have enough in their diet to pass on). The entire evolutionary point of milk is to allow infants to survive – and grow – in an environment where it couldn't do by other means and mobilising the resources locked away in mama's body is the way we do that. For humans, babies on the breast want 300–400mg of calcium a day, so mothers are supposed to be getting about 1,000–1,500mg themselves if they don't want to lose the stuff out of their own skeletons. That's 3–4 cups of skimmed milk or, because dietary calcium lurks in many foods and we might as well compare them, 3–4 cups of figs.* However, because it's actually quite difficult to eat 3–4 cups of figs a day, human mothers still lose bone to their babies. They just don't lose that much, probably less than 5 per cent. Milk is the miracle that allows you to take the delicious figs you ate in sensible amounts over the years, store up their calcium and then release it to build a baby at the correct time.

Even within one single species, the composition of milk changes. When a mammal infant is first born, there is no 'milk' at all and instead mothers secrete something called 'colostrum' – what we colloquially call 'first milk' if we call it anything at all. Colostrum is a thicker yellow liquid that mothers produce before 'normal' milk appears, packed full of immune-building, growth-signalling substances that essentially prepare a baby for a lifetime of eating and growing. However, it doesn't look very nice† and there's not much of it, so in a huge number of cultures around the world mothers are encouraged to delay breastfeeding until *after* this tailor-made immune-boosting superfood gives way to the white stuff. This includes a number of groups that some people might think of as living a more

---

* Note: this is too many figs to eat at once.
† It looks like pus.

'evolutionary' lifestyle, *i.e.* those who live by hunting and gathering or some other more 'traditional' subsistence practice than ordering food from a plastic speaker box while sat in your car. *

The medical authorities of antiquity were consistently on guard against the dangers of colostrum. We hear from one twentieth-century north London doctor with a history-of-medicine hobby that ancient Indian medical texts suggest a child should be given honey and clarified butter for these first few days before normal milk comes in, but perhaps even more surprising for those not of a certain generation, the doctor relates this to the 'modern' custom, handed down from 'the Greeks and Romans' of giving an infant sugar and water instead of letting them have the colostrum produced by the mother. This is, in fact, terrible medical advice, but seems to have been remarkably widespread both now and in antiquity. Katie Hinde has collected in a neat blog entry some of the very limited knowledge we have of different cultures' ideas about colostrum. The San people of southern Africa, who have a total no-farming way of life, share the same taboo against feeding a baby colostrum with thousands of years of Ayurvedic medical tradition from India and the dedicated farmers of the Navajo Nation in North America. Really quite a lot of cultures have taken against this first milk, either offering substandard alternatives (human or animal milk, gruels or other solids) or, rather more dangerously, nothing at all. One can only assume that humanity persists

---

* It is very important to clarify that all the people alive today have done exactly the same amount of evolution, because that is how life cycles and linear time work, so absolutely no one – no one – is living 'palaeo'. Unless you count whatever it is a bunch of very strict fitness enthusiasts are keen on. Whatever that is, it ain't old.

because the vast majority of mothers in the past did not always take the advice they were given, and those who did were well-buffered enough that they (and their babies) could take the hit.*

After the colostrum comes the deluge – the white stuff we think of as milk. But milk is not ever 'just' milk. It's not even the same milk it was when it woke up this morning. It contains different levels of nutrients depending on the *immediate* needs of the infant; in humans it changes according to time of day, weather, or how the baby (or mama) is feeling. The milk at the beginning of a feed isn't even the same as milk at the end of a feed. 'Foremilk', which comes down first, tends to be the high-sugar stuff, with more water content to rapidly get baby the sugar hit it is craving. The 'hindmilk', by contrast, has two to three times as much fat, the better to fill baby up. In hot weather, mothers produce more watery milk to protect baby against dehydration – something that zebras also do and another one of the reasons our milks might be similar. Milk at the beginning and end of the day is a lot lower in fat than in the afternoon. Morning milk is full of get-up-and-go cortisol, night-time milk has massively elevated levels of melatonin and tryptophan, the sleepy stuff commonly associated with eating too much turkey. Even the levels of immune-boosting and full-feeling factors in milk change over the course of the day. This 'chrononutrition' aspect of milk is something that researcher Jennifer Hahn-Holbrook and colleagues have pointed out has huge implications for the legions of women out there forced to rely on pumping milk. If a society was serious about caring for the next

---

* There is a comment to be made here about aspirational lifestyle websites and dangerous medical advice that I certainly will not be making in legally binding print form.

generation, this requires pretty serious re-evaluation of the economic systems that force a disconnect between when milk is produced and when it is consumed.*

Milk also changes as a baby grows. The very first milk we get is protective, full of immune boosting factors and has a nutritional profile that suits tiny tummies, but the composition of nutrients big and small is highly adaptive. For humans, we see that even though the overall energy carried in milk stays the same at about 70 calories per 100ml over the first year, the amount of galactose, nitrogen and, above all, protein decreases as time goes on. Protein is very important to growing babies, however, and as soon as they start to get their gums on 'real' foods the amount of protein in their diet starts to climb back up. Meanwhile, however, there is a steady flow of fat – and sugar. Human milk, like all primate milk, has some fat to it, but above all it is sugary; think multicoloured-milk-left-over-in-a-bowl-of-Lucky-Charms-sugary. We are actually born craving that sugar rush. This is why babies and young children have different taste perceptions of sugar, craving it far more than adolescents and adults who are no longer pumping masses of energy through their systems in order to grow.

We are actually so primed to like sugar that some clinicians use sucrose as a painkiller; many among us are of sufficient age to remember the lollipops doled out after a traumatic visit to the doctor.† This is not without consequence, of course; the same taste for multicoloured marshmallow-based foodstuffs that help a baby suffer through a blood test is also behind our growing health problem with obesity. The seeds

---

* I'll hold my breath, shall I.
† It is my understanding that children are now given stickers. Unlike sugar, there is no *Cochrane Review* data that suggests images of Peppa Pig function as an analgesic.

of our modern health crisis are sown deep in our evolutionary history, where we, like the rest of the mammals, have been encouraged by our perception of taste to chase the complex-carbohydrate dragon. Now, however, we are surrounded by easy sources of sugar, and no one has to wait for fruit season or climb a tree and fight bees for the hit. The early-life search for sugar has been augmented and encouraged to the point that our adult tastes are weighted towards the sweet end of the spectrum too.

Milk also changes according to what kind of baby you are giving it to. This is particularly fascinating, because it tells us something about the underlying investment strategy of the mothers doing the milk-producing. Who gets more milk, and who gets higher quality stuff, is a good indication of the line on which a species is putting down its genetic chips. If you recall the Trivers–Willard hypothesis in Chapter 6, the idea was that investing in sons was a shotgun strategy – a whole lot of shot with very little direction – while investing in daughters was more surefire. The lifetime reproductive output of a male has the capacity to get to Genghis Khan levels – a not-to-be-sniffed-at 16 million male descendants and counting – while the female is rather more constrained by how much of her physical being and time she can devote to building babies. So, what is a canny animal investor to do?

Well, a little unsurprisingly, they split the difference. According to studies that actually try to quantify how much milk is being produced in cows, daughters get more milk. Sons, however, get milk for longer. This might reflect the fact that you are going to get more descendants out of your lovingly crafted daughter if you get her plumped up to reproductive status more quickly, or it might be an after-effect of the hormones produced by different types of babies crossing the wires of the maternal milk-production unit.

This is the sort of thing you can study in cows, because there is an insane amount of money to be made from tweaking the slightest aspect of cow biology to produce more milk, and also because cows are trained to let you milk them. Most primates are rather less enthusiastic about being milked, bar, of course, the human mothers who can milk themselves and gamely volunteer for research projects. Hinde has managed to convince macaques to be at least marginally helpful in answering this question for the larger primate lineage, and her results suggest that mothers go big on sons when they are first born, giving them milk that is richer and less watery. But daughters get more milk overall, with more calcium in it. The major factors that affect milk quality then? Birth order, sex and how old baby is.

Oh, right, and what your society has evolved to tell you to do.

## CHAPTER ELEVEN

# Hey Diddle Diddle: the Cultural Life of Milk

Hey diddle, diddle,
The cat and the fiddle,
The cow jumped over the Moon,
The little dog laughed to see such fun,
And the dish ran away with the spoon.

You are what you eat, so the saying goes, but what it fails to mention is that humans have some very particular ideas about what babies ought to consume. Mammal babies need milk to grow, and this is true for humans just as much as it is for Pizza Rats, which is remarkable considering breastfeeding is *hard*. Even for those mothers who are able to stay at home and retrain as dairy workers, breastfeeding still sucks. Despite being absolutely critical to the species, most human women struggle to breastfeed at first. A good friend's mother is a lactation consultant, and having a consultancy in something is a pretty strong signal that people feel the need to consult about it. While not a situation that most non-reproductive people will have considered, the difficulty of getting something with absolutely no knowledge of the world to sort out feeding becomes absolutely all-consuming as an issue once you produce a squalling dependent offspring who wants milk *now* and also *always*.

Mothers suffer cracked and bleeding nipples, painful infections and swelling, even – as time goes on – bites.

And despite all that investment, the stripping of our skeletons, the pumping in bathroom stalls, the avoiding of caffeine and alcohol* – after about six months the nutritional quality of our milk starts to change, and the levels of protein needed to grow that baby have to be found outside our bodies and babies then have to learn about real food. We need to keep supplementing what infants can mush into their own tiny faces with some rich fatty sugary goodness, but there is a tipping point where the mother's needs and the baby's needs are no longer aligned – something has to give.

The second way to get yourself in trouble writing about human evolution† is to touch on the subject of breastfeeding, especially the evolutionarily correct duration and method of breastfeeding. There is a great deal of highly emotive opinion on the subject – and for very good reasons. Exclusive bottle feeding is now generally frowned upon,‡ for reasons that are partly to do with perceptions of the health benefits of breastfeeding and partly the result of a corporate scandal that made big waves in the late twentieth century. In the 1970s the charity War on Want published an explosive pamphlet exposing unethical marketing of infant formula in the developing world, titling it 'The Baby Killer'. One of the main objections to bottle feeding infants with artificial milk formula – and it was a big one – was that bottle feeding was cynically promoted as 'modern' and 'affluent', encouraging people to give up on breastfeeding and rely on the bottle, even in places where you

---

* With apologies to my offspring, not all mothers do this.

† The first is to start writing about human evolution at all.

‡ It approaches the status of mortal sin the closer you get to certain parts of north London, where I live.

could not guarantee hygiene or that mothers could afford enough formula to prevent malnutrition if their own milk dried up for lack of direct feeding. The idea that Western corporations were exploiting mothers' desires to do best for their child proved a rallying cry for breast-feeding proponents; for a corporation, promoting a method of infant feeding that was likely to carry dangerous, possibly fatal, bacteria, or lead to malnutrition, was not a good look.

Breastfeeding has several advantages over bottle feeding: there's the fact that it's cheap, available quickly and can be a pleasant bonding experience for the mother–infant pair. There is also evidence that those very first expressions from the breast after the baby is born, colostrum, are particularly virtuous, being high in nutrients and agents that help prepare the infant's immune system. That said, however, breastfeeding has had a difficult few millennia, if not longer. This may have quite a lot to do with the fact that breastfeeding has *been* difficult for probably several hundred thousand years. This is not to say all human breastfeeding is terrible, but to point out that despite our claim to be organic lifeforms adapted for mammalian reproduction, breastfeeding, like birth, is something our species seems remarkably bad at.

Modern breastfeeding statistics are pretty shocking. Exclusive breastfeeding, as recommended for the first six months of infant life by the World Health Organization, ranges dramatically across the world, from 0 per cent in Columbia to almost 90 per cent in Rwanda (see Figure 11.1). In the UK, approximately 1 per cent of mothers exclusively breastfeed to six months – the established WHO guideline. Only 34 per cent of mothers managed some breastfeeding until six months, while in

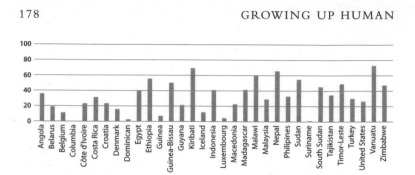

Figure 11.1. *Data on the percentage of infants exclusively breastfed around the world over the first six months of life, from the World Health Organisation (2019).*

the United States, nearly 58 per cent of mothers did. Statistics from other industrialised countries are occasionally better: China in the 1990s managed about 60 per cent of mothers breastfeeding some of the time until the baby was four to six months, while the rate of exclusive breastfeeding hovered around 30 per cent; though this seems to have risen since.

Given the range of these figures, it seems safe to say only certain groups of humans are terrible at breastfeeding. But why is this, given that breastfeeding is something we have shared all the way back to our not-quite-a-tree-shrew days? One factor might be in how we approach breastfeeding to begin with. It is only in the past few decades in Western medical practice that skin-to-skin contact between the mother and newborn within the first hour of life has been recognised as potentially important for triggering the reflexes in both parties that stimulate breastfeeding. Originally trialled as part of a 'kangaroo care' solution to providing underweight or premature babies with a beneficial environment, initiating skin-to-skin contact in any size or gestational-age newborn has now become a default practice in many delivery rooms.* A recent *Cochrane Review* – the gold standard in medical metastudies – wasn't conclusive, but it may be that hospital practices in industrialised countries were interfering with setting up successful breastfeeding.

That is one explanation for the massive achievement gap in global breastfeeding, but there is another that has the weight of reams and reams of evidence behind it: differences in social and economic status. In countries that are less economically well-off, the percentage of breastfed infants is higher. By contrast, in higher income countries, breastfeeding rates at a year or more of age are almost nil, while they are still around 80–90 per cent in the lower- or middle-income countries. The opportunity cost of breastfeeding seems to favour continuing on in lower- and middle-income countries across the world, but there are sharp deviations when you look at the richer or poorer parts of individual societies. These deviations follow opportunities for women to enter the labour market, and to negotiate time away from it without suffering adverse consequences. Poorer mothers in poorer countries breastfeed for longer than their affluent fellow citizens. Richer mothers in richer countries breastfeed for longer – or at least they do since about the 1960s. How much we breastfeed isn't so much about being good or bad at it – it's about whether we have the chance to do it. In low-income countries where infant formula is *still* marketed as aspirational, there could be good social reasons for upwardly mobile mothers to choose to formula feed; in wealthy countries where the time and opportunity to breastfeed is an unthinkable luxury, the same

---

* If you are in the USA, skin-to-skin contact is often an optional extra. One Utah parent was charged $39.95 for the privilege, receipted alongside the rest of their medical care and, perhaps inevitably, posted to twitter. https://twitter.com/benkling/status/1103717087808290816

is true for breastfeeding. With humans there is a cultural factor determining what should be a standard biological process.

There is one final consideration when it comes to seeing the differences between our modern practices and what most humans would have experienced prior to the invention of alternative infant-feeding techniques like formula feeding, and this is that we seem to need to learn how to do it. It is not obvious, nor is it as instinctual as the fluffier end of the internet might suggest. Where other mammals just get on with things, primates need to learn how to take advantage of what we are told is one of our last inbuilt reflexes. It's not just humans – research has shown macaques struggling and many monkeys are just, quite frankly, very bad at babies the first time around. But the rather cruel thing is that human mothers are made to feel particularly bad about it. Having the capacity to rationalise is perhaps not ideal in facing a biological process women are told is 'natural', without having the support or advice necessary to position their own experience; ethnographies of first-time mothers in the UK have suggested guilt, stress and concern about pain were all factors in struggling to breastfeed when they are told it 'should be the most natural thing in the world'.

Controversially, I suggest that perhaps we should consider not shaming women who don't manage to breastfeed. There are factors beyond individual ability or grit at work in societies that do achieve a high rate of breastfeeding, such as the act being a normal public activity, with social support and advice from experienced mothers readily available. Mammal milk maven Katie Hinde and anthropologist Brooke Scelza were able to discuss breastfeeding with Himba mothers from Namibia, where every single mother was able to breastfeed, something that certainly isn't true in most

Westernised cultures. The authors picked out two key factors that accounted for this remarkable success rate: the support of other mothers, and the absence of any taboo or stigma associated with breastfeeding in public. The things that inhibit breastfeeding in other societies – like working hours or social rules that prevent frequent feeding – are simply not a part of Himba life, along with the stress or even the guilt associated with not breastfeeding.

Despite the fact that breastfeeding has been part of our mammalian history since the littlest weasel-like milk producer peeked out from behind a stegosaurus, it is not actually that easy for humans, or for human babies. Breastfeeding is a skill, and it is one you can learn easily in a community where it occurs out in the open and less easily when you don't see much of either the act itself or the experienced mothers (like your own) who can help out. But since it is unlikely that the rest of the world is going to immediately reorganise itself into a mother-friendly environment perhaps, especially in contexts where baby weight gain can be monitored and high-quality infant formula used safely and hygienically, we could leave off worrying that we are not meeting some evolutionary ideal and haranguing mothers who, for whatever reason, aren't breastfeeding. The end goal of human evolutionary adaptation, after all, is to avoid dying. Everything else is just a bonus.

Parents in the past were just as keen as we are today to make sure their children were fed. Given the difficulty of breastfeeding and the many circumstances that could mean a mother wasn't able to feed her child directly, we would imagine baby bottles to be of considerable antiquity. So, it is a bit surprising that we do not seem to see them much in the archaeological record. We have to assume that before we see the very few examples that come down to us in clay and metal, there would have been organic flasks and vessels of horn or skin or gourd or wood or any of the other lovely organics that leave little to no trace for us to find. And, sure enough, once we invent pottery, there are some very interesting ceramic beakers that start to appear in central European baby graves from around 7,000–8,000 years ago. This also happens to also be right about when we, as a species, got into dairy. These vessels have little spouts that could have been used for feeding a baby and sometimes even come in adorable animal shapes – like the odd little 3,000-year-old bunny-eared, cotton-tailed jug found at the site of Franzhausen-Kokoron in Germany. Chemical analysis of the trace residues rolling around the bottoms of the vessels that uses a kind of acid reaction to identify lipids – fats – found that they had contained a substance with high fat content. Vaporising the residue to determine its elemental structure – more properly known as gas chromatography – revealed that the triacylglycerols in the fats were much more characteristic of dairy fat than, say, barbecue grease. These, then, are indeed ancient baby bottles and they are joined by many others – the little spouted vessels found in two 3,000-year-old infant graves at Jebel Moya in Sudan, the one shaped like a pointy-toed shoe and decorated with a smiling face from Pouzzoli in Roman Italy, and many more besides.

Unfortunately for the owners of the German feeding bottles, they may not have been enough to keep infants alive in the past. As discussed in the previous chapter, human milk has very specific nutritional properties and the milk of our domesticates isn't quite up to the same level as the real human thing. Adding to the danger of

Figure 11.2. *Line drawing of a baby feeding bottle from Vösendorf, Austria, c. 1000 BC. Based on an image used with the kind permission of Dr Katharina Rebay-Salisbury.*

bottle feeding in the past is the ever-present threat of disease-causing bacteria to little immune systems; the several-hundred-dollar boutique bottle-sterilisation systems available to the modern un-financially-constrained parent were not exactly a thing in Germany in 1000 BC.

That these vessels are found in the graves of children does suggest that those children did not flourish. For parents of the past – lacking modern medicine and the vast array of nutritious substitutes for human milk – feeding an infant anything other than breast milk for the first six months had the potential for disaster. Many mothers struggled to produce milk, something we can see in very early texts: from the anonymous physician authors of the 3,500-year-old Egyptian 'Ebers' Papyrus, to Simio San of the 2,500-year-old Qianjinfang text, to Avicenna 1,000 years ago, authors throughout history have discussed the trials and tribulations of feeding a baby. Advice and proscriptions abound. A rather extraordinary* account of breastfeeding throughout human history by the odd London doctor from the last chapter, for instance, tells us that Egyptian women struggling with milk production ought to have swordfish bones heated in oil rubbed on their backs.

These bottles and texts form the front line of defence against the strange way we view infant feeding, that in the past we must have done it *properly*; that no self-respecting Upper Palaeolithic mama was going to let her infant down by not feeding them directly from the breast at all times, for years and years longer than we do today, with nothing but fully organic produce and mammoth purée to follow. This is giving the practice of infant feeding the 'Palaeodiet' treatment, where the past is imagined as Edenic and everything is 'natural'. While it is certainly true that before animal domestication, obtaining alternative food for a very small baby by, say, creeping up on an unsuspecting lactating onager and milking it is going to be very difficult, not to mention profoundly unwise, there is also the fairly certain fact that milk production can fail, and that Upper Palaeolithic parents would have wanted their babies to live just as much as any parents.

---

* The level of casual racism in Ian Wickes' ethnographic descriptions of breastfeeding data makes for stomach-roiling reading; he imparts that certain races have pendulous breasts that allow them to flip one up over their shoulder to feed an infant on their back (this is a bizarre invention of racist eighteenth-century Dutch colonisers in Africa, not an actual thing). He also tsks as only the wealthy mid-century English physician can at the terrible mixing of 'primitive' and 'modern' habits seen in a four-year-old who is pictured enjoying a cigarette after breastfeeding in a country that is not England. Never forget where some of our 'science' comes from.

There is every likelihood that before complacent milking animals became a thing, humans would have stepped in to share milk, much like the 'milk bank' services that operate around major hospitals today. In many cultures it's not odd at all for women to share milk with unrelated babies; this can be part of an important cultural bond that ties their families together. I can think of several examples from among friends and colleagues in Turkey that follow an old 'milk brother' custom, where two otherwise unrelated babies are brought into a close social relationship by being fed by the same woman. Not all milk-sharing arrangements are created equal, however, and with a stratified social system in full swing, the urban world is exactly the place where you might find people willing to leverage their biology for material security. This applies to breastfeeding for money just as well as it does any other physical transaction.

The word we use for this is 'wet nursing', when the job of breastfeeding a baby is handed out to someone who is not the baby in question's actual mother. Unlike the cooperative milk banks there is a far more mercenary edge to how we understand this role in the past, especially as it starts to get written into history in the form of actual contracted employment. Our first textual glimpse of wet nursing, which no doubt lagged after the first experience of milk sharing by several hundred thousand years, comes from Mesopotamia, when contracts for hiring a wet nurse were first scratched into clay. We see in the Laws of Eshnunna, a sort of 4,000-year-old precursor to the more popular Code of Hammurabi, plenty of mention of wet nurses and their contracts, and throughout the period early texts called cuneiform tablets reveal the minutiae of the arrangement, e.g. ten shekels to be paid for three years of breastfeeding and raising a child.* It became de rigueur for the wealthier classes of antiquity to outsource their breastfeeding. While this might have been motivated by the inconvenience and potential discomfort involved, it may also have been prompted by the desire of wealthy women to return to fertility more quickly.

Whatever the case, for thousands of years the royalty of south-west Asia and the Mediterranean hired wet nurses. Some became like second mothers, travelling with their charges into adulthood, like Princess Bēltum of Qatna, who took her 'ummu' with her to the royal court of Mari when she married, more than 3,000 years ago. In dynastic Egypt, wet nursing could be an extremely canny path to social elevation. Some 2,500 years ago, the rather extraordinary female pharaoh Hatshepsut saw to it that her nurse Sitra was given a burial in Egypt's Valley of the Kings, normally the preserve of royalty. Things change, however: 1,500 years later and the high-status women of Egypt wouldn't dream of nursing a baby. A papyrus letter sent by an irate mother-in-law to one Rufinus, her daughter's husband, insists that he hire a wet nurse – she is not going to let any daughter of *hers* nurse.

Down the line, the history of wet nursing has continued to the modern day, though different social meanings have been attached along the way. While medieval European royalty would have shared Rufinus' mother-in-law's horror at the idea of breastfeeding their own infants, by the early modern era wet nursing had got so

---

* Failure to pay at the end of the contracted period may result in your child being confiscated and sold. The small print on Mesopotamian law codes is really *very* small, but quite important.

out of control there were near-constant scandals that more or less ended the practice out in the open for Europeans. The ideal in recent history, as captured in newspaper ads posted in London in the late eighteenth century, was that a sober, respectable woman would have somehow just become widowed (and therefore not be having another baby) and also would be possessed of a good supply of milk.*

Wet nursing was sufficiently lucrative, and the cities of the last several hundred years so polluted and dangerous, that sending children out to be nursed frequently meant actually sending them out. A nurse in the countryside could provide the kind of bucolic health anxious urban middle-class London mothers dreamed about and babies were indeed sent out in droves to the countryside, where they would be visited on occasion, but largely left to grow in a 'healthier' environment. However, the balance between commercial success and the actual physical ability of a woman to provide adequate milk was a delicate one. Human milk has changeable nutritional content throughout the period of lactation; provision adapts to the age of the baby and all sorts of factors. The thousands of 'nurse-child' babies recorded in the burial registers of the parishes surrounding London in the seventeenth and eighteenth centuries must reflect some of the risk involved.

And, of course, not all wet-nursing arrangements were so equitable. What on one hand could be a socially – or materially – beneficial exchange could also be a deeply

---

* How many of these husbands were fictitious is hard to say – the number of ads claiming the nurse's husband had 'just gone off to East India' is rather striking.

exploitative one. The compulsory diversion of a woman's biological resources towards another woman's child is well known from antiquity; slaves in Greece might be expected to nurse their mistress' child, even before their own. It would be nice to be able to say that this was a phenomenon restricted to the ancient world, where we can imagine all manner of cruelties at a safe, dehumanising distance, but there are harrowing accounts in the very recent past of enslaved women in the Americas being forced to feed another's children, while their own offspring suffer and possibly die.

Feeding a human infant has other corollaries, of course. I refer here to the several-hundred-thousand-year experience of getting our extremely altricial, helpless babies not to poo or pee all over themselves, their parents and other annoying-to-clean surfaces. This long and seemingly unwinnable battle has been fought by parents presumably since time immemorial, but sorting out elimination takes on a particular importance when you do not move around much and toileting is a much more fixed experience. Changes in the ways humans live – settled, indoors – necessitate some ingenuity in baby-butt management. By the time you reach urban population densities, you need to make sure your entire society has sorted out something to do with waste, baby-produced or otherwise, or you are not going to have that population density for very long. Urban life necessitates a series of intermediary steps between the individual and waste disposal. For the urban child, acquiring these rather critical skills will ensure a level of hygiene that (hopefully) will keep your society alive – and you in it. In the modern world, we have a number of adult cultural options for going to the bathroom: there are methods involving water, paper, squatting, sitting, nearby greenery, *etc.*

For babies we only have two competing philosophies: pants or pants-free.* Given how generally undesirable it would be to curate baby diapers, it is not particularly surprising that we have very little material evidence for them. What we do have is the occasional visual evidence that people had already made the calculations of relative effort in terms of diapering: in a culture that practised swaddling fairly religiously, some bright spark in ancient Greece had the idea to wrap up a baby but leave a gap at the back, clearly visible in a little 2,400-year-old votive figurine found in the harbour town of Piraeus that looks totally swaddled in cloth from the front, but when flipped over clearly has its behind hanging out.†

Around 2,500 years ago, nearby Athens was a bustling metropolis full of parents, babies and one particularly oddly shaped bit of pottery that went to ground in the famous Agora, only to be dug up again thousands of years later in the 1930s, where it would proceed to baffle the learned men who tried to describe it. This pot resembled two open bowls, stuck with their bottoms together – but the bottoms had been cut out, so the thing was more like a drum with a pinched waist. It would have stood a little over 1ft (0.343m) high, and a skilled artist had painted on it plants,

---

* The Chinese tradition of split trousers, where toddler age children are dressed in trousers with no middle bits to get in the way of immediate evacuation, are something of a middle ground. Though you would think they could be a hazard in crowded areas, and are apparently not much used in cities nowadays, I have certainly sat next to infants so attired on the train and emerged from the experience utterly unscathed.
† The prominent archaeologist Kristina Killgrove has written about several of these items in her hugely enjoyable and accessible column for *Forbes* magazine, and I am indebted to her for the introduction to pants-free Greek baby swaddling styles.

Figure 11.3. *Researchers tested their interpretation of the ancient Athenian potty chair with the help of an anonymous child in a nappy who was unable to use the potty successfully (left) and Elizabeth Carlyle Camp (right), without nappy, who was successful. Image of object P 10800 courtesy of the Athenian Agora Excavations and the Ephorate of Antiquities of Athens City, Ancient Agora, ASCSA: Agora Excavations © Hellenic Ministry of Culture and Sports/Hellenic Organization of Cultural Resources Development (H.O.C.RE.D.)*

animals and mythical figures − geese, a lion, two rather friendly looking sirens. The top 'bowl' had a rather wide opening and there were two big handles on either side for lifting. The excavator, a man experiencing a degree of nominative determinism called Homer Thompson, could make head nor tail of it. It would take decades for a classicist from the British Museum called Peter Corbett to examine this funny, if pretty, shape and work out what it was: an ancient Greek training potty. Dorothy Burr, an archaeologist who had obtained her doctorate from Bryn Mawr in 1931 and had become the first female fellow on the giant excavation of the Athenian Agora that revealed the potty – she also subsequently married the aforementioned Homer – noted that the potty was just about the right size to fit the bottom half of a posh ancient Greek toddler.

Where there is one potty, there are bound to be more – and so it proved. Subsequent study and re-identification of otherwise baffling sherds of pot have yielded up numerous baby potties in the Greek and Roman worlds; an occupied one is even depicted on a lovely black figure vase, with proud mama watching on. While these substantial pieces of baby kit were likely to be limited to the more affluent reaches of society, we can still see them as important pieces of evidence for baby care. So many other physical aspects of baby and toddler life were likely to be woven, carved or sewn from organic materials that it is always fascinating when we do find something that links us directly to these tiny little lives in the past. Lucky finds in the Alaskan climate have discovered, for instance, baby pants made out of seal skin by a culture related to the modern-day Iñupiat. We need to have perfect preservation conditions to retrieve such ephemeral bits of human lives as clothes or blankets, so it is always something of a miracle to find little patterned leather shoes from ancient Rome because it's been lost down an anaerobic sewer for thousands of years, or the sandals of a young child buried 2,700 years ago in the semi-mummified graves of Tiwanakuan Peru, or the startling burgundy cloth and cashmere hat preserved in the burials of a baby mummified in the arid conditions of the Tarim Basin in north-eastern China.

When we consider the cultural adaptations of our species to the business of baby care and feeding, the truth of the matter is, if you are reading this book, it probably doesn't matter much whether you are a baby-wearer, a co-sleeper or a long-term breastfeeder. The most important intervention possible to make in a child's life has already been made – by raising you, the parent, to a level of socio-economic status where you do things like purchase quality non-fiction books about important evolutionary topics. This status goes along with a number of other corollaries – access to healthcare, the social and financial support needed to educate you, the leisure time to read – that are all associated with improved outcomes for baby. It is almost impossible to detach the importance of class and access to resources from the life chances of an infant. Even the largest-scale medical studies are beholden to the groups of people who participate in them, and the way humans live cannot be separated from their ability to meet their material needs. *This* is why it is difficult to parse the actual risks associated with different infant care practices. The risks of bottle feeding a lower-income child in a country without access to safe water are very real. The risks to lifetime achievement of a bottle-fed child in an economically comfortable and educated household with a nice dishwasher that runs at sterilisation temperatures? Probably minimal.

And this is the thing to remember when we talk about childhood in our long peripatetic past and try to work out what we have 'evolved' to do. A Palaeolithic childhood was adapted to living in the Palaeolithic age. Some of it might indeed be very applicable to our modern lives – children still require feeding, for instance – but some things may have changed markedly, like the foods we have available to feed them. Does this mean our children are doomed to sub-optimal lives if we cannot source organic blended mammoth paste? Of course not. But the appeal of palaeo-parenting remains while we, as a species, are confronted with the staggering uncertainty of how to do what is best for a child. Meanwhile, the world's largest economy, the USA, has no provision for women to take time off for the business of producing new citizens. Very few places on this earth where women earn wages are in any way set up to accommodate the actual biological business of building babies – and even then there are clear divisions in terms of access to support. As an academic at a major research institution in the UK, I had the option[*] of 18 weeks' full pay on the successful production of a baby. If I had been working as an outsourced, zero-hours contracted, completely disposable employee? Six weeks at 90 per cent of pay.[†]

At six weeks, a human baby wants feeding every two hours. Little babies aren't even that good at feeding; some of them need a good 30–40 minutes every feed; pumping isn't that expedient either, even though you can now rent hospital-grade industrial pumps.[‡] Going back to work, which might not be a choice for the economically precarious – something that is already more likely to describe women than men – usually means something has to give. The number of women gamely trying to produce milk in toilets and broom closets around the world just to give their kids what they believe to be the best start in life is absolutely galling. In cultures where mothers' economic beings are not physically separated from their social and biological existences, feeding babies is less of a big deal. Carrying your baby with you – a sling will do even when you haven't got the body hair for a

---

[*] In the UK, either parent can take the paid portion of parental leave. So, actually, I had two weeks' paid leave.
[†] It took concentrated strike action and a heck of a lot of hue and cry for the 'outsourced' staff at my institution to be given the same rights as the rest of us, but it did happen – very recently.
[‡] Because that will fit in your purse.

baby to cling to, or a baby that is any good at clinging, for that matter – and the normalisation of breastfeeding as something unremarkable and done in public contribute to a near 100 per cent rate of breastfeeding in some places in the world. But, for some reason, we don't seem to think that this particular investment is worth it. Maybe if more people understood milk, those scales would balance a bit easier.

## CHAPTER TWELVE

# Papa's Gonna Buy You a Mockingbird: the Evolution of Dads

Hush, little baby don't say a word,
Papa's gonna buy you a mockingbird,
And if that mockingbird won't sing,
Papa's gonna buy you a diamond ring.

We, cleverest of the primates, have found ways to help provision our kids above and beyond what a single human might be able to do. But who else might be entrusted with investment in babies? There are a few options that might spring to mind, and whichever first appears in your head probably says a bit more about the society or family you grew up in than strict evolutionary biology. For instance, if none of your responses were 'Dad', you might have a hard time conceptualising paternal investment within the X-chromosome-heavy cycle of reproduction. This might be because the same structures of your society – power, capital, biological determinism – that allow you to purchase quality non-fiction books have inculcated you with an idea of child-rearing that is very female-centred.*

You are not, however, alone, if you didn't immediately come up with the reproductive partner as your response to

---

* Or, it might be because you're reading this in the future timeline where gendered parental roles sound absolutely medieval, in which case, hey, how's that time machine coming?

the question 'who cares for the baby'. Research into parental investment is no more free from the biases than actual human beings are.* It is, however, remarkable that early anthropology took so long to consider whether the way we think about essential human nature might be a bit coloured by the societies we live in, given that is the point of the whole study-of-humans business, but there you go. We had to get through a lot of man-the-hunter-type 'descent of man' theory before boomeranging all the way right out past theories of a prelapsarian unified Mother Goddess religion to reach a more nuanced approach to reading the past. Early approaches to the study of primate parenting were no better. Darwin himself was utterly scathing on the subject of the value of human females, judging sexual selection in evolution to have resulted in 'man's attaining to a higher eminence, in whatever he takes up, than can woman – whether requiring deep thought, reason, or imagination, or merely the use of the senses and hands'.†

Susan Sperling writes eloquently about the phenomenon of 'baboons with briefcases', our very human tendency towards anthropomorphising the behaviour of other primates and the difficulty for researchers in extracting themselves from the perspectives of the society that produced them. There is a specific way of looking at the world that produced ideas of male competition – as embodied in the over-the-top fangs of the male baboon – as

---

* Do read Angela Saini's *Inferior* and *Superior* if you want to know how bad scientific biases can be.
† Darwin, C. 1871. *The Descent of Man*. He also allows that women are almost OK, but in a fantastically racist way: 'With woman the powers of intuition, of rapid perception, and perhaps of imitation, are more strongly marked than in man; but some, at least, of these faculties are characteristic of the lower races, and therefore of a past and lower state of civilisation.'

the ultimate driver of evolutionary prowess, just as there is a worldview that sees female-led cooperation in chimp life as a major factor in a primate group's success. It may be that setting biological sex as a black-or-white factor in human evolution is not as helpful as we think. Often, Sperling argues, behaviour we think of as either purely male or purely female is posited as having some huge adaptive significance, only for us to realise a bit later that perhaps the behaviour wasn't all that sexually determined. She gives a great example that might impress Pizza Rat: rodent males that are around newborns can end up displaying female behaviours, including things like the rather non-adaptive yoga 'table pose' female rodents adopt for nursing. Care, even the kind that seems to be biologically hardwired into sexually dichotomous categories, may be a more adaptable behaviour than the evolutionary theory of yore assumed.

Dads are just one example of a very important primate strategy: 'cooperative breeding'. There are two ways to raise offspring:* either the mother can go it alone, investing heavily in each offspring and bearing the brunt of reproductive costs; or cooperatively, allowing other group members to aid her and aiding others in her turn, collectivising the risks and rewards. In many social animals we see a great deal of what is called 'alloparenting' or, more frequently, 'allomothering' – animals caring for infants that are not their own. Alloparents could be the baby's siblings, aunties, uncles, other relatives or just the mother's friends – all are attested and all are valuable in the societies (human and non-) that have them. Not all alloparents are created equal, however. For starkly hierarchical species like the ubiquitous macaques, there is a real danger that

---

* Correctly, or incorrectly, according to your grandmother.

higher-ranking females, enchanted by new babies, will kidnap them from subordinate females, even though they aren't lactating and have no way to feed them. If the lower-ranking female can't get her baby back, the infant is liable to serious injury and possibly fatal neglect from its new caregiver. By and large, however, alloparental care is beneficial, sharing the burden of care and feeding. In most mammals, alloparenting is a pretty female-centred business, but in primates like us we start to see an increasing reliance on one particular family member: Dad.

The primacy of primate papas is a question of evolutionary adaptation – and when those patterns of care change, as they do in humans and some monkeys, we should pay particular attention because these are the strategies that got us to where we are now.* The most famous encapsulation of how different parental investment strategies might manifest in animals comes from the work of yes, Trivers again.† He argued that parental investment is a limiting strategy for reproductive success and whichever parent was putting the most into the offspring they already had wasn't benefiting by having more offspring elsewhere. For mammals, a limiting factor is our ridiculous pregnancies and period of breastfeeding, so females are automatically stuck with the role of the investing parent. The question for males, strategy-wise, becomes whether they can offer

---

* Jack of all adaptations, master of none.
† Trivers had many *cough* insights, among them an offhand comment that human females attractive in adolescence 'tend to marry up' and the rest essentially get into heavy petting because that's what they've got to offer. This is why a) you should probably get an anthropologist to check your evolutionary biology and b) the 1970s were such a miserable place for women in anthropology (or indeed, women in general).

enough help building superior offspring that it's worth sticking around.

One of the major hurdles in understanding the evolution of fatherhood in our species is that we actually aren't entirely sure about it. We, as educated persons reading this book,* are aware that paternity uncertainty is a major concern in many human societies. It underlies several core rules covering female virginity, marriage and adultery, all of which can get you into different levels of trouble depending on whether you are transgressing social norms in a free-love commune in the LA foothills in the 1960s or somewhere a bit more straight-laced, like the seventeenth-century Puritan colony at Massachusetts Bay.† Despite the somewhat racist assumptions to the contrary by early anthropologists, most notably articulated by the august 'father' of the art of ethnography himself, Bronisław Malinowski, humans *are* generally aware that sex leads to babies and of what that implies for fatherhood.‡ More critically for anthropologists is the question of whether this knowledge of paternity – or just the fact of it, if it's a bit beyond the macaque paygrade – is something that actually underlies how we invest in our children.

---

* Or, more honestly, as participants in a culture where infidelity (and discussion, dramatisation and *dénouements* of same) is a major theme of entertainment and/or public life.

† While the commune sounds like fun and execution less so, they both end badly. At least they didn't have Charles Manson in colonial Massachusetts.

‡ The Trobriand Islanders Malinowski studied may not have *cared* about paternity in the same way as a man from a patriarchal land-holding society where reproductive success is determined by the kind of capital you get from paternal inheritance ... but that doesn't mean they didn't know what fathers were.

We know what 'paternity uncertainty' means in our species, particularly in our societies that have rigid social institutions for ensuring that paternity is pretty damn certain. But as we discussed previously, there's no clear mechanism for *how* animals that do not burn people at the stake for adultery and have not invented genetic testing would work out which infants to give the benefit of fathering to. Without a specific mechanism to act on, it's hard for paternal investment to be seen as something that can be selected for and emphasised in order to have an evolutionary benefit. Indeed, most animals that don't do the large-gamete part of reproduction have opted to spend their investment on free-living free loving – searching and competing for mates takes up their energy, rather than caring for offspring.

Among the well-known champion animal fathers are the seahorse dad that lugs his fertilised eggs around, or the clownfish father whose inept but well-meant parenting skills are immortalised in *Finding Nemo*; birds like the wandering albatross father that does more than half of the job of flying literally thousands of miles to find food for his chick – and a surprising handful of primates.[*] Now, the seahorse knows those are his fertilised eggs – he's carrying them. And albatrosses have an absolute dedication to their pair bonds. It is only a small subset of mammals that show some paternal investment – something like 5 per cent of the total. But why – and how – would primate fathers, who are mostly pretty promiscuous, start to invest in their children at the expense of a – cough – broader social experience?

---

[*] Honourable mention here to the Swedish phenomenon of the 'latte papa', who spends a small fortune on ruggedly good-looking childcare accessories and lattes to drink while wearing them during his vastly generous paternal leave.

As we saw in Chapter 3, the social role of dads is part and parcel of the social organisation of the rest of their species. These rules not only determine who gets to mate, but what involvement those mates then have with their offspring. If you have to spend all your time and energy growing big enough to fight rivals for the chance to mate, or seeing off marauding infanticidal invading males, it's going to drastically cut down on your energy for baby care. While primates overall show more paternal involvement with offspring than mammals in general – about 10 per cent of primates do some sort of direct paternal care – it seems that most primate males still prefer to do a lot of their investing around conception. This makes sense because most primates live in polygamous or promiscuous mating systems where time and energy are spent competing for mates and then guarding the ones you've got. So, are dads just what you get when you get monogamy?

Well, no. Monogamous primates, or groups of related females for that matter, might be understood to have a stronger motivation for alloparenting than the more promiscuous-male species, but in fact parental care shows up across a very wide range of primates. Just because you get around doesn't mean you don't love your kids. There is evidence of polygamous primate fathers investing preferentially in their offspring, even in promiscuous social systems; baboon fathers, for instance, will usually back their sons in a fight. Chimpanzees are even more confusing – chimpanzee dads are nicer to their own babies, but it's not hugely clear whether that's because it's adaptive, or because they're trying to get in good with the mother (which might be just as adaptive). Chimpanzee sons for their part spend quite a lot of time grooming their dads, though they may never know that they are their dads – it may be that males that are nice to chimpanzee

202	GROWING UP HUMAN

sons' mothers are just a safer bet in the dangerous and potentially violent world of adolescent male socialisation. Protection from infanticide is another major investment in infant care, and one that several researchers have suggested might be an early motivator for primate dads to take a closer interest in their offspring – and take a more monogamous approach to mating to be certain the kid they're protecting is theirs.

Not killing your offspring, however, falls a bit short of what we humans might consider adequate parental care.* Primatologists split paternal care into two categories – direct care, like actually feeding or holding the baby, and indirect care, like making sure it doesn't get eaten by a lion. Primate dads are in general better on the lion front than the holding-the-baby front, but that doesn't mean that *all* primates are deadbeat dads. For some of the smaller and furrier New World monkeys, a caring dad is the only way to survive, because their partners reliably produce exactly one more baby than is practical for them to carry alone. The dinky but handsome tamarins and marmosets of the New World have evolved to release multiple eggs, meaning that twins are their default setting – and that means you need Dad around to carry the baby. Titi monkey dads are the primate champions in this, carrying their babies something like 90 per cent of the time, with the result that the infants become distressed if separated from their fathers, but not so much their mothers.

Pair-bonded, monogamous living is one way to offer the primate father some sort of security in knowing it's his own genetic line he's investing in, every time he carries that baby. Without the threat of marauding hordes of males

---

* Yes, yes, I know, not in your grandad's day. Uphill both ways and no grooming.

trying to displace your genes through infanticide, monogamous primates are able to add considerably to their babies' chances in life – which, from a genetic point of view, means their own chances of reproductive success. Pair-bonded living isn't a guarantee of fatherly care, however, nor is it necessarily tied in to your place on the primate family tree. Siamangs, for instance, will tote their offspring around once they're about a year old, but they are the only gibbon to do so, despite gibbons generally being fairly monogamous.

Prolific alloparenting in pair-bonding marmosets includes both males and females – many of whom might be related and so benefit in an auntie-like way from the continuation of the mother's genetic line. But it's not always so straightforward. Despite being classified as 'monogamous', literally no one knows what's going on in the world of marmoset parentage without a DNA test. For marmoset dads, it might be that genetic lines, plural, are what they are investing in. Due to some extra-curricular mating on the mother's part and a very weird placenta, marmoset babies can be genetically 'chimeric', *i.e.* made of different fathers in different parts. Chimera babies seem to get more attention from the pool of potential dads around, suggesting multiple dads is yet another adaptive primate strategy.

With the notable exception of our New World baby-carrying monkeys, most primates are raised exclusively by their mother until a certain age, after which they begin exploring the world with the help of friendly group members. While this absent *paterfamilias* role practised by most primates might sound reminiscent of a certain archaic 'have the child brought to me once it can speak ancient Greek' ethos, it is not the universal case in our species.

What about human fathers? We sit on the very far end of the spectrum of paternal care for mammals. Actually,

we sit on *both* ends of the spectrum. Depending on the age
of the child and the social rules of the society, human
fathers are incredibly involved with their offspring, or not
involved at all. There is of course the case of indirect care:
one study estimated that 97 per cent of calories eaten in
ten different foraging societies were collected by males.
But, additionally, there is a lot of direct care. A survey of
186 cultures observed over the last hundred years or so
found that 40 per cent had an active and involved role for
fathers and, given that ethnologists of that era had a
predisposition towards categorising all people and things
with rather bloody-minded rigidity, this measure of 'role
of father' may miss quite a lot of nuance.* More detailed
studies have identified the wide variety of roles available
to fathers in human societies, from the Ba'Aka† of Central
Africa who carry their infants something like 22 per cent
of the day, to the American dads who struggle to get 50
minutes in. What human fathers definitely do have is the
potential to improve their offspring's chances. In some
cases, two fathers are even better – societies that allow for
a loose identification of dads can turn up a few extras, and
a study among the foraging Barí people of Colombia found
that kids who had multiple proposed dads were 16 per cent
more likely to thrive.

---

* In the original publication, there is quite a lot of statistical ritual
applied to declaring dads to be more important in societies where
descent isn't patrilineal, sexual relationships are monogamous or
polygynous, boys aren't circumcised, there is no overarching 'boss
God', adolescent boys are socialised in the main group and where
games of physical skill but not strategy are played. This is ... overly
restrictive. Playing chess doesn't make you a bad dad.
† Foraging societies that live in mostly forested areas around the
Congo river are also described as Aka, BaYaka, Baika, Ba'Aka, Baka
or, quite pejoratively, 'pygmies'.

Fatherhood, just like motherhood, is heralded by actual chemical changes. Just like our determined little male rodent adopting a nursing pose, the presence of babies messes with the internal chemistry of human males. Just like in women, oxytocin, the so-called 'happiness hormone',[*] rises around babies. Oxytocin, cortisol (associated with stress response) and prolactin (the 'milk' hormone – in females, this promotes lactation) all rise depending on the amount of association with the pregnant partner before birth. The main focus of research on changes to dads, however, has been on testosterone. This is the world-famous sex hormone that, in males, is linked to many different sorts of growth – particularly changes post-puberty – and (mostly, in animals) to behaviour.[†] Generally, testosterone is lower in human males not competing for mates – something we might think of as linked to being in a pair bond – something that may be adaptive for species like ours that live in huge multi-male and multi-female groups, saving men the hassle of expending their efforts on mate competition. And one of the behaviours accompanied by changes in testosterone levels? Parenting. High testosterone has been observed in males when they are competing for mates from mice to men; those levels drop by up to nearly 40 per cent in human men who stick around after the baby is born. Caring activities like hauling an infant around has a similar, though not as extreme, effect – particularly in primates like in our baby-wearing

---

[*] So-called because it is not; there is no more a hormone for happiness than there is an absolutely monogamous primate.

[†] It is not, despite considerable popular assertion to the contrary, securely linked to just being a jerk. Testosterone levels are weakly linked to how humans respond to being challenged, but how you deal with that car cutting you off on the freeway is all on you, buddy.

marmosets. If the father is *not* doing much parenting, testosterone rebounds pretty much as soon as he's away from the pregnant lady.

This brings us back to the critical question: how did we get these useful fathers? None of our nearest relatives have them – none of the other great apes are getting on the podium for Father of the Year any time soon. Given the highly varied locations in our primate family tree that paternal care pops up, it's actually quite difficult to isolate the adaptive evolutionary value that made *our* species choose to embrace fatherhood. The arguments provided for the evolution of paternal care in primates and other mammals may not specifically apply to us – we may be monogamous, but we are group living (and we don't shut off the reproductive capacity of subordinate females), so it is unlikely that paternal care evolved for us in exactly the same way it did for marmosets and tamarins. But we may not be so unlike the seahorse or Nemo's dad where fatherhood is the ultimate extra investment that sees babies to successful adulthood.

It's of course possible that fathering is only sometimes a human behaviour. We see both kinds of males in our species: the pair-bonded paternal carer and the promiscuous deadbeat.\* Perhaps we have evolved to simply monopolise whatever strategy works best in each circumstance – it would certainly be on form for our species. But we do have dads – and dads do important jobs, ones that are critical to the success of their offspring. Jobs that require investment, above and beyond a few cocktails and possibly some cheese fries. The key to raising a successful baby – one that goes on to have other successful babies – is all of that post-birth

---

\* And it's not like they have banded tails or marked ruffs so you can tell them apart, like the different types of dads in the lemur family.

investment that is so critical to all species, not just our own, and dads are a major part of that story. If you want to be human, you've got, in the immortal words of Kurt Vonnegut, to be *kind*. You've got to care. We spend the first part of our lives utterly dependent on our parents for the feeding, carrying and general care that allows our species to get from its stupid, dangerous altricial state into something that is ready to learn to be human. And that is the thing that we have really messed with in our quest for world domination: bringing up our big-brained babies.

## CHAPTER THIRTEEN

# There Was an Old Woman Who Lived in a Shoe: Lots of Babies – Fast

There was an old woman who lived in a shoe.
She had so many children, she didn't know what to do.
She gave them some broth without any bread;
She whipp'd all their bums, and sent them to bed.

Despite the increasingly ominous tone of several reactionary politicians in calling for women to breed like there is no tomorrow (for the propagation of their particular ethnic, linguistic or cultural group, somehow forgetting all of these traits are reproduced culturally and not biologically), dropping endless litters à la Pizza Rat is not, in fact, the evolutionary strategy that made us great. Reproductive success, it turns out, is not about how many babies you have. It's about how many successful babies you have. Quality over quantity is the slow-life-history mantra. But even so, we have a *lot* of babies. In this chapter we will have to reconcile two opposing but powerful adaptive truths: we are simultaneously slow-life-history animals, but yet wonderful at reproducing ourselves fast, fast, fast. How do we explain this contradiction? What is it about human childhood that leaves room for so many of us?

This is one of the most fundamental questions we can ask about the unique nature of our species and, to my mind, a dramatically unconsidered one. Population growth is a bugbear of Malthusian pessimists and evolutionary biologists while biological reproduction is the realm of medicine; trying to understand the interaction of these two rather separate fields in the human past is the occupation of palaeodemography, which starts off with the unreasonable handicap that everyone it has ever cared about is dead. There is, however, considerable insight available from taking what we have learned about the physical processes of human reproduction and situating it within the ecological system in which they occur: basically, thinking of ourselves as a primate on a mission. How, with all the obstacles we face, have we become the most successfully breeding primate on the planet?

Let us start with the story of our success – and we are definitely a success. Humans have a lot of babies, compared with other apes. Bar certain realms of cinematic fantasy, this has left us with unquestioned dominion over our little planet. But we are too big, and born too unfinished, to ever qualify as fast-living animals. So how have we managed to go slow, but still coat the planet with our offspring? Humans do actually have the potential for a very rapid rate of reproduction, something we haven't always managed to take advantage of due to the availability of resources,* but will not be unfamiliar to people with recent family histories of grandparents (or parents) who were one of a number sufficiently large that it can be rounded into terms normally reserved for the sale of eggs. There are nigh on 8 billion of us, while our closest comparators, the great apes, struggle to stay viable with populations of a few hundred thousand, or even less in the case of the severely threatened orangutan. We exist in these numbers because we found a way to increase our populations way beyond that of any other primate species, and there are exactly two

_____

* Or suitable mates. Ask the Neanderthals.

ways to exist in extraordinary numbers: you either live forever or you have babies faster. We have gone for menu option two – and it is indeed a menu option, because how we feed our babies and for how long has a direct impact on how many of them we have.

The age at which a baby stops draining its mother of milk is one of the primary dimensions of our human childhood and something that we can see is strongly linked to adaptive behaviour. As we saw in earlier chapters, all mammal milk is not created equal, and the way it is doled out has a real impact on the shape – particularly the length – of childhood. 'Parking' mothers like galagos need to deliver more nutritious milk in less time, and may be motivated to get that baby grown and moving on its own as soon as possible, so as not to make of it an owl snack. Similarly, the 'ride or die' primates are protecting their offspring from predators by staying very, very nearby while the infant is vulnerable, and might be able to afford to let baby be a baby for a little while longer. For these longer-term baby-carriers, there is an additional cost to be faced. It is far easier to trust arboreal childcare than to constantly haul an actual infant around, slowing down the animal giving the ride and burning up their valuable calories.* It is in the interest of the galago to have a baby that grows up fast because clingier primates can stay infants for longer – and their parents suffer for it. Hard choices have to be made that balance the time it takes to get a baby from birth to independence against the species' need to churn out more babies.

Just like the time it takes to gestate a baby (as we saw in Chapter 6), the time from birth to being kicked off the

---

* And, presumably, patience.

breast to make way for a new baby is linked to both the overall size of the animal you want to grow and to how much growing you need it to do during this stage of life. So, looking at the big-eyed tarsier, the primate answer to the question 'are there too many lizards?', we can see that it is born at a whopping 20–30 per cent of its total eventual size, because it needs to be able to get on with killing lizards quickly or it will starve. Compare the tarsier experience to another primate that spends six months gestating: the capuchin monkey, who is only about 10 per cent of its total adult body mass at birth. Capuchin babies are not leaping about on their own at the age of two months, they are clinging to their mothers as they have been from birth. They are bigger, and they are pretty good at clinging, but they are far less developed at birth than our tarsier – it's going to take them almost a year and a half to stop drinking milk and become independent of their mama. Gorilla babies are on the breast until they are about three years old. Chimpanzee infancy can extend up to six years, though usually babies are weaned, or put off the breast, somewhere between four and six. The same goes for the solitary orangutan, weaned at around four years old, but occasionally nursing well beyond that to seven or eight years. Our nearest primate relatives have infancies that stretch out just about forever, or certainly what seems like forever if you, a human, are in the midst of dealing with an infant of your own. The timing of transitioning an infant off of breast milk is an absolute key moment in primate life history; not only is it a big step for baby, but it's a massive leap for mama (and her potential to go off and have more babies).

There is an extraordinary and frequently quite snippy debate in circles anthropological, medical and 'parenting-advice-website-ical' on the subject of how long humans should breastfeed – and even how long they actually do breastfeed. We really do not actually know how long we are 'evolved' to breastfeed for. There is reasonable ethnographic evidence that Inuit-speaking groups in the Arctic breastfed children on occasion until seven years of age. The WHO recommends breastfeeding to two years and beyond, but today we applaud mothers in the most developed countries on Earth who manage even six months.* It is a really remarkable thing to not know when your species stops breastfeeding. It is a key marker of how fast we get back in the baby game. Breastfeeding undermines the ability to get pregnant again, by throwing the maternal body into a hormonal soup from which (for humans) it takes about six months to climb back out of. This is called lactational amenorrhoea: when you don't have reproductive cycles, aka periods, aka the chance for more babies.

It has been canonical for a long, long time to say that human groups that subsist on hunting and gathering live on this knife edge of nutritional sufficiency, so can use this period of no periods as a type of birth control. There is certainly considerable ethnographic data saying that women in foraging groups in the past century or so tend to have babies spaced much further apart, with three to four years between births. Agricultural communities, by contrast are thought to reproduce much faster, with gaps of only a year or two between babies, judging by both historical data on births and direct observation in the modern day. There is only one problem with this neat story: it's not actually about the food. The mechanism that supresses ovulation – and

---

* You get better at it, and some mothers adore breastfeeding, but oh the joy when they start to feed themselves.

periods – is a complex hormonal signalling that is triggered not by a return to optimal fat, but by the simple act of suckling. In humans, a small study suggested that eight suckles a day keeps the periods away, no matter how much actual energy is transferred between the mother and the baby.*

This means that the factor suppressing fertility is actually behavioural – the ability to have another baby is determined by the frequency of nursing; the very act of parking or riding is what shapes how long our infants are infants for and gives us the first bend in the strange arc of our childhood. Of course, there is still potential for the role of nutrition to affect things, because it might be that an undernourished mother produces less satisfying milk and has to feed more often, for longer, to achieve the same level of baby growth. Body mass index (along with age) is just about the only thing that changes the fatty content of milk.

As we saw before, lifestyle and milk production are strongly interlinked, with opportunity costs that vary depending on how fast you need your baby to become an efficient lizard-killing machine or how long you are able to tolerate your kid literally hanging off you. This has big implications for when you can have your next baby. Long-nursing orangutans, very sadly given the massive environmental challenges they face, are slow reproducers, with a between-baby spacing of almost eight years. Chimpanzee females have an interbirth interval of around six years, while gorillas manage closer to four and a half. But

---

* You can absolutely 100 per cent get pregnant while breastfeeding.

humans? Properly energised, we can work up to one a year. And *that* is how you take over a planet.*

For several decades, the accepted orthodoxy on the 'invention' of village life, known to archaeologists as the Neolithic period – has been twofold. First, people settled down because of a rapid cooling shift in climate between about 12,000–14,000 years ago, called the Younger Dryas, which made foraging a less secure way of staying alive than exploiting unmoving plant resources. We have good evidence for this climatic change – and equally strong evidence of changes in what kind of game was available to hunt – and the changing environments we hunted in, so this is less controversial. Secondly, however, comes the idea that settling down is the root cause of the major expansion in humankind that will eventually engulf the entire planet. This theory is known as the Neolithic Demographic Transition, and has been taught to several generations of researchers as established fact. I myself have uncritically used the shorthand of this theory to explain changes in the human lives I have researched. But, like all good theories, the idea of the Neolithic demographic transition needs to be poked vigorously with a scientific stick, to see just how well it holds water when the facts come stabbing in.

The theory as originally proposed by palaeodemographer Jean Bocquet-Appel starts with a simple premise. An increase in births in a population, while all else remains stable, should lead to population growth. Likewise, a decrease in deaths. And in the Neolithic, what we see is a baby boom. Technically, rather, it is a baby bust – there is a sharp increase in the number of infant and child burials associated with early farming communities, which is how we actually know

---

* Also, cultural transmission of knowledge *etc.*, but let's save that for another book.

there were more babies in the first place. More babies dying, a little counter-intuitively, suggests a massive uptick in population fertility because unless the proportion of babies surviving changes, it must mean there were more babies to start with. So here we come to two arguments that need to be upheld for us to believe that the big swell in human populations is due to year-round residence and an increased reliance on plant foods. Either of the twin pillars of Neolithic society – farming or sedentarism – need to be shown to increase human fertility. At the same time, we need to know that the record we find of these skeletons that are thousands and thousands of years old isn't biased somehow, and what we are seeing is a real uptick in human fertility that is fuelling population growth.

Does settling down increase human fertility? There is certainly strong ethnographic evidence to suggest that changes in lifestyle can send birth rates soaring. One of the classically cited cases comes from the Aché people of Paraguay who had lived a foraging lifestyle until being more or less forcibly settled in the 1970s. Prior to being settled, Aché parents spaced their children about three or four years apart. Their livelihoods depended on a fair amount of movement and one can see the disadvantages to trying to move through the rainforest with more than one toddler at a time. If you contemplate, for a moment, the wisdom of taking a toddler *and* a baby into the rainforest and hoping to come back with something accomplished, or indeed hoping to come back at all, you will start to see some of the reasons our behaviour – and our birth spacing – might change.[*] Rather mechanistic research

---

[*] This group at least avoided the murder, rape, and torture visited on other Aché people by Paraguayan ranchers who hunted them for sport and took them as slaves.

among women of the San people who live across the central southernmost African nation states worked out that carrying more than one child would 'de-optimise' their foraging strategy.* After switching to a more settled life, birth spacing dropped to closer to two years. This precipitous drop in birth spacing and resulting increase in baby making is exactly the kind of baby boom imagined for the Neolithic, the point in human history where some of us became settled farming people; a shift in activity that leaves open the possibility of more alloparental care 'back at the ranch' and a rise in mothers' nutrition due to farmed foods that fuel fertility.

The four-year birth spacing originally observed in the Aché equates well with birth spacing in other modern groups that make their living as foragers and hunters. In recent foraging groups like the Hiwi of the Orinoco Basin in Venezuela, or even the San, a drop in nutritional status means a drop in fertility and a lengthening of the time it takes to have another baby. Increasing the space between children lowers the overall birth rate. Not all foraging groups have long interbirth intervals and low growth, however. The Hadza of Tanzania space their babies a little less than three years apart, while the Baka of Central Africa have children about every two and a half to three years. If you were to stick to long birth spacing, you would get low or no population growth, which is more or less what demographers have spent decades insisting we had back in the day. But birth spacing is determined by a range of factors, particularly in humans who have a range of behavioural options available to prevent pregnancy. The idea that something like the San pattern of fertility is

---

* Lovely name for a child, De-optimise.

'typical', with four-year birth intervals and no population growth, has been seriously challenged, including by the original researcher, and new information on external causes of infertility like venereal disease have come to light.

Humans have a whole slew of cultural adaptations that define our behaviour and thus our reproductive habits. In addition to staying put, which allows for the potential of the safe containment of toddlers, we have incredibly strong rules and regulations on who ought to marry whom and when, and these have far more effect on our societies than straight-up nutrition. Despite everyone's favourite Shakespearean love story featuring barely pubescent protagonists,* the average age of female reproduction is much later than the age at which she is *able* to have a baby, particularly in developed economies where women do wage labour. Delayed mother-hood isn't just a modern phenomenon either. While armchair sociologists† are very fond of assuming humans have always reproduced as early as possible in order to get those numbers up, neither the historical record nor ethnographic data hold this up. Country by country, the average age of mothers when they have their first baby is around 24, with the lower end of that range at around 20 years old and the higher almost a decade later. And this later start in motherhood isn't just a modern thing. Historically in the UK, for instance, the average age of marriage (which usually preceded the first birth by a year, though by all means not always) for rural women of modest means was the same in the 1770s as it was in the 1970s. Careers, particularly in domestic service in the

---

* If you don't read *Romeo and Juliet* as a teenager the entire concept sours, and by the time you are 40 you want to shout at them that it's just hormones and they'll get over it. Or maybe it is timeless literary art. One of the two.
† And anthropologists who should know better.

eighteenth century, delayed marriage to an average age of 26 and held there until the twenty-first century. Today, the average age of marriage for women has skyrocketed to 35, if they get married at all.

We care less about men in the story of timing infancy because they haven't got the gestational time constraints that women are up against, but men do reproduce slightly later, depending on cultural traditions and economic conditions. So even though we could have babies earlier, mostly we choose not to. And even if we could have a baby every year, we mostly choose not to. We also do one more very remarkable thing, that almost no other animal does. We, alone among great apes, tend to outlive our fertility. This puts a cap on our numbers, and yet we have still managed to out-populate any of our fellow primates. These are all rather counter-intuitive life-history decisions, but they are the ones we have made and that have made us. The challenge comes in understanding how and why.

Perhaps one of the most readable bodies of work on this subject comes from Sarah Hrdy, whose volume *The Woman That Never Evolved*[*] had the temerity, in 1981, to add the consideration of females to the study of primate reproductive social behaviour. While Hrdy was of course not the first to observe female primate behaviour, she did offer a reframing of it as an important part of the construction of the social underpinnings of reproductive success. One of the most salient points made in the volume is the

---

[*] There is something entirely apropos that the copy of this important feminist volume held by my institution, University College London, was donated by the archaeologist Juliet Clutton-Brock − whose career at the Natural History Museum, where I have had a long association, included analysis of the animal remains from my research site on the island of Chios. Apropos, or, possibly, terribly insular.

attention received by outrageous male behaviour – aggression, murder, infanticide – may outweigh its actual
evolutionary importance. While male strategies for reproduction appear to consist of both making a bloody mess
and captivating primatologists, the more subtle strategies
of the female have often been overlooked. Some strategies
aren't even that subtle. While the rest of our primate relatives have been merrily living out the spans dictated them
by sensible models of energy in, energy out, and fitting
appropriate periods of dependence and maturity into their
size-appropriate life spans, humans have been busy developing a world-changing evolutionary female adaptation.
Something so radical that it may be single-handedly
responsible for our reproductive success and the subsequent overrunning of the planet; something so unlikely
so as to be almost unbelievable: *grandmas*.

Depending on your personal experience of grandmothers,
the fact that they are a radical force in human evolution
will not come as a surprise.* And radical they are. In
Chapter 2, we discussed how different life histories lead to
different periods of dependence – how animal 'childhoods'
might be determined by the strategies a species uses to
survive, whether they live 'fast' or 'slow' and how much
investment each offspring requires. The models of how a
species might spend its allotted span of days rely on an
evolutionary trade-off between long, drawn-out periods of
immaturity that result in greater adult size and greater adult
capacity for reproduction *or* a quicker march towards
adulthood to get all that reproduction in before you die. So
far, so clear. Except nowhere in this model is there room

---

* This is especially true if your grandmother taught you the words to
The Weavers' 1949 progressive anthem 'The Hammer Song' as a
small child.

for life *after* adulthood's reproductive function. There's dependence, then there's growth, then there's reproduction and death. Yet here we are, awash in females who are not reproductive and *still not dead*.

The idea that grandmothers are a major benefit to humankind will not, I think, shock most people. There is a collective ideal of the grandmother as a warm, nurturing presence that is only occasionally supplanted by the actual experience of a woman who has lived through several more decades and probably children than your actual parents, walked uphill to school both ways in the snow and has *no* time for your malarkey. Culturally, we accept grandmothers as part and parcel of our family units; a Jungian archetype built on top of the scaffolding of a maternal figure by adding on some quaint old-timey exclamations, lies about hill geography and a few wisps of grey hair. Grandmothers seem natural to us, in the strange ecologically determinist sense of 'nature' that we use when we want to describe something that *is*, without ever questioning *how* it got to be.*

Well, let evolutionary science help you out with thinking about how you came to have grandmas: they are, in a biological sense, very, very suspicious. Humans are some of the only animals in the world whose females will live a long while past the age at which they can normally reproduce successfully. Grandmothers, animals that have lived past the ability to reproduce, are hugely unlikely. An incredibly small number of animals live past the ability to reproduce – even our own species' males maintain some

---

* 90,000 words on the subject of people who use the word 'unnatural' to describe human relationships that don't adhere to some Victorian notion of the nuclear family available upon request, or upon submission of one and one half strong gin and tonics.

generative capacity until nearly the end of their lives. Here
are some animals with grandmothers that are done having
babies: Orcas, short-finned pilot whales, us. Here are some
animals without grandmothers: every other animal in the
world.* There are plenty of species where some animals
don't reproduce for whatever reason – perhaps they are not
the dominant breeding pair, perhaps their mate has
died – and continue to live without reproducing, but this is
a very different thing to what we (and those whales) do,
which is to live *past* our biological capacity to reproduce.
Now, many animals do have a period of senescence, or old
age. No one wishes to take the position of matriarch away
from the venerable old lady elephants of the African
savannah, but for many animals reproduction is still a
possibility right up until the end. Even a 70-year-old
elephant still has viable potential-baby-elephant eggs left in
reserve.

In humans, however, there are actual physical changes to
the whole body, which, some time around the fifth decade,
tell the reproductive system to pack it in. The accepted
theory of reproductive biology for at least 30 years tells us
that, like all mammals, human females are born with the
total number of eggs – technically, oocytes – they're ever
going to have – something like 2 million of them. This is
actually a climb-down from the peak number of nearly 7
million while baby is still growing in utero, mid-gestation.
The vast majority break down without ever having had a
shot at release; by puberty there are only about 400,000
proto-eggs left that are capable of turning into a baby.
About a thousand more disappear each menstrual cycle,

---

* There may be other animals, such as the chimpanzee and some
whale species, that have post-reproductive females, but it's going to
be a list you can count on your hands.

some of them contributing to the hormonal soup that launches one specially selected proto-egg each month towards its shot at personhood.

Losses, even at this scale, would be predicted to still leave the human female capable of reproducing for 70 years – if something in our biology didn't interfere to shut the whole system down, because by the time we get to that fifth decade, human women are officially out of eggs and out of the baby business. For those of us who waited perhaps a little long for family life* it is mildly infuriating to know that human females don't have to run out of eggs. New research has even offered tantalising hints that our birth eggs may not be our destiny. Mice, it seems, can recruit cells into being oocytes well into adulthood; similar cells have been found in humans. Whether or not we can recruit new eggs, the current operational facts on the ground remain the same however: after about 50, we either stop making them or we run out, but either way we are done.†

It turns out, we are not the only primates who check out on child-bearing around 50. Our nearest living relatives, the chimpanzees, also stop reproducing around the same time, even though they still have reproductive cycles. Research has shown that they run out of eggs at a roughly similar rate to humans until about the age of 35. After 35, though, it seems like the chimpanzee march towards the end of reproductive life carries on at a different, slower

---

* *I.e.* those who may have foolishly pursued years of higher education and then misspent the prime reproductive years shunting around temporary academic positions, travelling out to dig sites, living in tents and generally having a terrific time without the benefit of a recognisable adult life. Or wage.

† Fertility research, however, is certainly not 'done'; it might be some day we figure out how to turn that egg-recruiting system back on.

pace than that of humans. Our cadence sees us rapidly turning off the tap in terms of more offspring, but we do OK on other measures of senescence – that's ageing – and can expect decades of healthy life after the end of childbearing. Chimpanzees, whose allotted years number about 50 in the wild, but can stretch much closer to human spans of more than 70 years in captivity,* also generally stop having babies some time in their fourth decade, and may actually also experience something like menopause in terms of measureable hormonal changes – something that has only recently been observed in chimpanzees living in relatively luxurious environments that allow for longer life spans. The oldest chimpanzee mother on record was an august 56 at the time of the birth and a study of long-lived captive chimpanzees showed that most continued their normal reproductive cycling until death, or very shortly before.

While the record for oldest human mother is constantly being extended, thanks to the advances in fertility treatments, less than 2 per cent of babies born around the world are born to mothers over 45.† At around age 50, the vast majority of human women will go through menopause – defined as a full year without menstrual cycling – and lose the capacity to reproduce well before the rest of the natural ageing processes catches up with them. Meanwhile, human females also have life expectancies that

---

* The current record holder is 'Little Mama', a chimpanzee born before the Second World War, whose death in Florida at the age of 79 was noted by obituaries and tributes from around the world, and no doubt set a high-water mark for chimpanzee retirement-planning.
† The oldest allegedly non-IVF conception is reportedly that of Tian Xinju of Zhudong, in Shandong province, China, who swears she gave birth at age 67 after relying on traditional Chinese medicine. You'll have to take her word for it.

extend beyond those of even the best-loved chimpanzees; farther, even, than the males of our species. If you were to take the simple life-history equations that we use for other species (as seen in Chapter 2), which extrapolate life span from body mass, humans in general actually outlive all expectations. We outlive our bigger relatives the gorillas, we outlive our massively bigger and not-very-close relatives the elephants; we outlive every other mammal besides serene and august seafarers like the blue, fin and bowhead whales. In other words, we may shut down the reproductive part of our lives, firmly and promptly, but we make up for it at the other end. The question then becomes: why?

Something, somewhere, at some time has selected for longevity in humans. Evolutionary pressures have shaped us into that mythical animal the Sphinx is always asking people about – we go on four legs, then two, then three as we crawl, stride and hobble through our allotted span of years. But in females, as we've just discussed, a lot of those years aren't spent reproducing, which is the thing that evolution is supposed to pressure us most about.* So we must look for another payoff, one that still measures success in reproductive terms, but uses a lever other than the simple 'having more babies'. It is in that space we find grandmas.

This is the Grandmother Hypothesis, to give it its full title, proposed by Kristen Hawkes and colleagues in a 1989 paper that emerged from their observation of family life among the Hadza. The authors proposed that grandmothers contributed to reproductive fitness of their genetic line by bringing resources to their offspring with young children, taking some of the burden of providing for a grandchild on themselves so that their daughters might have the reserves

---

* A pressure visible *in extremis* as that applied by aspiring grandparents on their adult children.

to reproduce successfully – and more often. A Hadza grandmother is an excellent source of the tubers that contribute the majority of calories to the camp's diet. She can provide additional food, not to mention acquired child-rearing wisdom, a pair of hands and general moral support that, if given to her daughters with small children, directly benefits her genetic relations. This is a clear benefit – the grandmother isn't motivated to hoard those resources herself for her direct offspring since her own kids should be mostly grown by the time she hits menopause; instead, she can benefit the next generation.

This, then, is our secret weapon. Care. From grand-mothers; from dads and aunties and uncles. These are the ways we provision a baby that is too big, too helpless and needs too much. If we map it out against all of the live-slow and live-fast creatures of our world, we see we have moved the slider bars that mark out the stages of our lives to allow for a fast gestation and an undercooked baby. And now, we have to move them again, because we're not done messing. We have clawed back some time from the other end, shutting down reproduction early, creating the utterly unlikely phenomenon of grandmas, but it's *still* not enough to grow a human. We need more. More time and more investment. And that is the story of our most fascinating evolutionary adaptation – the long, long intermission between the business of being babies and the business of making them. You and I call it childhood.

## CHAPTER FOURTEEN

# The Mouse Ran Up the Clock: the Long Primate Childhood

Hickory, dickory, dock.
The mouse ran up the clock.
The clock struck one,
The mouse ran down,
Hickory, dickory, dock.

At the start of this book, we looked at the life-history strategies an animal might choose to pursue – fast or slow, 'r' or 'K'. But this is big-picture stuff and life is not lived at the macro level all the time – there are no species-level committee meetings discussing adaptive strategies.[*] Life histories are made up of smaller sub-stages, milestones on the path to evolutionary success – the way we grow can be described in little short hops as well. When and how we set up these milestones reflects both the environments we live in – it is hard to grow very big if you live somewhere very small – and the behavioural adjustments we have made, like trading faster growth and denser milk for a less demanding and more portable child for longer. All of the 'choices' animals make are trade-offs between the

---

[*] This may not be entirely true for humans – I am writing in the spring of 2020 in the near-global lockdown only reluctantly adopted by my government due to the emergence of a novel coronavirus and there have in fact been actual committee-level meetings about how to adapt to a lethal pandemic. Apparently 'herd immunity' is not only a biological phenomenon forced on species by an evolutionary death match with a virus but also a fun, cool new trend you can actively choose to participate in.

competing costs of growth (and avoiding not-growth, aka death) and reproduction; these choices are what determine the position of the slider bars of different stages of life.

We have seen that gestation is geared towards growing a certain type of baby, and that different babies need different levels of milk and attention during the critical stage of infancy. But what about the rest of it? There is a long haul between being a baby and being a baby-maker, and that is the space where we fit our childhoods. And just like how long we cook our babies for, or how long we keep them on the breast, childhood has very different dimensions depending on what you want to do with it. We have already seen that humans have cut short their infancy, and possibly even their gestation, for the size and nature of the animal that we grow, but when it comes to the period that comes next – childhood – ours has been extended beyond all recognition.* Nothing is as well known in the field of primate childhood as the diagram by A. H. Schultz that illustrates the stages of childhood in ourselves and our nearest living relatives, and from this you can clearly see where humans are stretching the limits. Despite being the same size – and considerably lighter – than some of our primate relatives, we grow up slowly.

Observing primates is all very well, but there is an entire realm of hard biological data just dying to tell us about the nature of human childhood. Literally – it is the stuff we ourselves are made of – our skeletons and teeth – and this fascinating archive is what biological anthropologists like myself are consulting as we recover the bones of primates past. We can chart where a body is on its path towards adulthood using the shape and appearance of the

---

* My mother still buys my shoes.

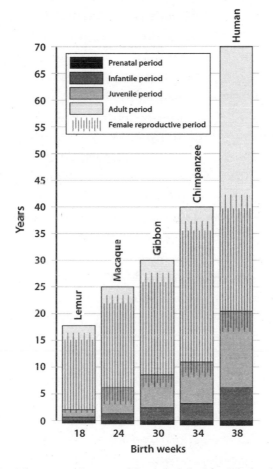

Figure 14.1. *The stages of primate life as described by Schultz (1969).*

skeleton – and because growth follows such a familiar trajectory, we can place it on a developmental timeline. This developmental timeline is, of course, the stuff that life history is made of and comprises all the pieces of our strange childhood.

Live fast, die young, leave a good-looking corpse – this is the strategy of an animal that gestates quickly, reaches food independence quickly and gets straight to reproduction. The doting dodderers, by contrast, invest more energy in their offspring than can possibly be transferred on a tight deadline. In humans, the landmarks of gestation, the length of time we feed directly from our mother, and the subsequent growing-up phase we identify as childhood and then follow up with a sort of halfway house to maturity we might as well call adolescence, are all spread out along an increasingly long lifespan. And we know this, because these evolutionary choices are written directly into our growing bones. Growth is a line that describes a species trajectory from conception to death and this is what we are really talking about when we discuss what our bones look like.

The skeleton is there to do a job and the way we grow it reflects what it needs to do. It needs to get big, because babies start small, and it has to be a certain shape and mass, because of gravity, but we aren't lizard-people,* forever enlarging in all directions – we have a set adult size where growth finishes. This is because growing big is only one part of the species mission – we also need to grow more people, so we start to shut down growth in our skeleton when we hit our reproductive stride, even if that stride is way off that of our primate cousins. The milestones of life history – gestation, infancy, childhood and the beginning of our adult reproductive careers – all leave traces, literally, in our bodies. It is the job of the anthropologist to identify these traces, and to build a story of growth that we can compare not only to our nearest and dearest primate relatives, but also to the rest of the animal kingdom, pulling out hard physical evidence of where we put our effort, and investment, when it comes to becoming us.

---

* Lizards grow their whole lives, but despite certain popular conspiracy theories to the contrary, there is no evidence that any actual lizard-people exist.

So, what evidence tells us about our bizarre childhood? There are many factors that affect how fast the skeleton of any given creature will grow, or at what age it will fuse various pieces of itself into their adult form. This 'fusion' is actually pretty critical – immature bones have a characteristic unfinished appearance, like little sticks or plates, and all of the complicated end bits start off separate. The way to ascertain the phase of growth is by judging their size and shape, and the end of growth is clear because the bits all fuse together. A big animal needs a bigger skeleton and a bigger skeleton will take longer to grow – *ergo*, big animals should have a longer period of growth: a longer period where their skeletons are still immature and unfinished. There are also the laws of physics to deal with – an elephant *needs* more skeleton than a mouse, and is made up of about 27 per cent skeleton by weight compared with the mouse's 4–5 per cent because, as prominent anthropologist B. Holly Smith wryly points out, an elephant with the skeletal proportions of a mouse would be a structural disaster.

Growth itself is a trade-off: having immature bones, ones that are not quite finished and so are a bit soft or a bit in pieces still, is fine for a while, but may not be the most stable solution to moving your mass around. We find that certain animals prioritise maturing and stabilising certain areas first, depending on what they need their skeletons to do, meaning they harden up their bones and solder together pieces of bone that start off loose in different areas depending on how they use them. What we need our skeleton to do acts as a check on how much we can grow – those elephants need a better bulwark against gravity than mice, so their bones solder themselves together far quicker than the mice, which actually scamper so lightly through their entire span on earth they never bother to completely

knit their skeletons together.* We can see where variation
in maturity is able to give us an insight into what an animal
is using its body for and that is something we have definitely
changed over the long course of our own species' evolution.

Primates don't mature their skeletons the same way
most mammals do, which people have suggested is very
much a reflection of the odd ways primates move around
in the world. You have vertical climbers and leapers like
the acrobatic tarsiers and lemurs, who fling themselves
through arboreal environments with a nice balance of
ankle strength and wrist.† Then there are the quadrupeds –
rangers of the savannah like baboons and even quite
limber climbers like macaques fall into this category, and
they can be seen making their way on all fours, with fists
all curled up or fingers spread out depending on whether
they are happier on the ground or in the trees. Brachiators
are the proper Tarzans, straight-backed and swinging
from branch to branch, with outrageously powerful arms
stuck out to their sides for maximum shoulder rotation
and power. This shows up in the little apes, gibbons and
such, and seems to have arisen independently in spider
monkeys. But, interestingly, even the apes that are far too
big to hurl themselves between branches – all the great
apes – share the signature straight-backed brachiator build

---

* Though the same can be said for marsupials, and kangaroos aren't
exactly light. A zooarchaeologist friend once suggested this as a basis
for positing marsupials as the key to eternal life, but I am reasonably
sure she was joking.
† This can be over-adapted, as in the case of the sifaka, which is
capable of leaping 40ft but has overcompensated in the leg
department. If you ever wish to see an animal that moves almost
exactly like a Jim Henson puppet, find a video clip of a sifaka trying
to get anywhere on the ground.

that would not be out of place at your grandmother's dining room table.*

Is movement what makes our skeletons? Well, matching the pattern of skeletal growth to the pattern of movement hasn't been entirely successful. Mammals generally fuse and finalise the bones of their ankles first, then hips then knees, but primates have rejigged this order to stabilise first the hips, then the ankles, then the knees. This may reflect the funny ways we move, or it might just reflect a funny way an ancestor moved and we got stuck with it. In the upper limbs, many primates follow the established mammal pattern of elbow-wrist-shoulder, but some − notably a handful of New World monkeys, chimps and gorillas − go for stabilising first the elbow, then the shoulder and finally the wrist. For some researchers, this has suggested that flying through the trees with the greatest of ease was sufficiently critical to species success way back when in the primate lineage that how we grow actually changed to accommodate strong brachiating arms. Like most other primates, humans sort out their elbows first, then wrists, then shoulders. But chimps and gorillas, which have shoulders that suggest they are at best ambivalent about grooving through the treetops, actually fuse their elbows then their shoulders, and afterwards the wrists, something that has been theorised to reflect their specialist adaptations to life on the ground. So, there you go − there are no easy answers.

What seems to be the most critical thing is to time finishing off our skeletons − fusing them − right, so that our skeleton is ready to withstand the maximum possible

---

* Even the tree-centric orangs don't brachiate − their locomotion is described as a 'quadrumanous scramble'. It looks even less dignified than it sounds, which is presumably why they do it so slowly.

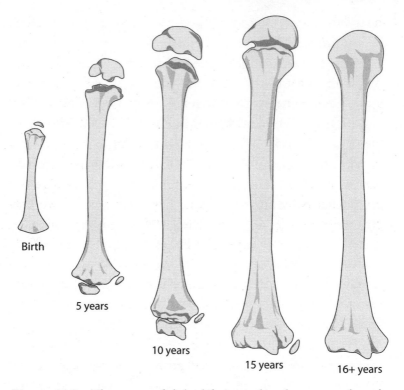

Figure 14.2. *The process of skeletal fusion in long bones maps how the diaphysis (or shaft) and epiphyses (or end-parts) grow and fuse together.*

force when we need it. Research shows that young chimpanzees 'finalise' their skeletons just before becoming primarily quadrupeds, so right before they adopt normal adult movements after testing out a wide range of fairly humorous methods of locomotion as infants and juveniles. The moment at which a baby is able to move around by itself is an important life-history milestone: locomotor independence, in anthro-speak, is a turning point in how much energy parents have to expend keeping their baby alive. After the intense investment in gestation and infancy, and the truly demanding period of trying to keep an incompetent wriggling child fed, a child who is able to

move around the world on their own is a significant relief.[*]
The point at which a baby monkey achieves this independence is pretty strongly linked to the speed at which they
grow and what they are growing: longer limbs for a bigger
animal take longer to put together in working fashion.

And not only do we need time to grow legs so that we
can wander off and be children, we also have the extra
burden of growing those pesky costly brains, as we saw
earlier. Balancing the energy needs there – what mama can
give, what baby needs – is what determines when we kick
off childhood. Where we have data, it seems like primates
prefer to sort out moving just before they come off the
breast, so that independent moving and the skeletons that
support that are strongly linked. This is great news for
those of us who study skeletons, giving us a clue, in the
shape of fused or unfused bones – as to where on the
developmental timeline a previous primate was. However,
the bigger primates have worked out how to push this to
the limit and can buffer some risky extra size – and
brain – growth by staying on a mother-supplemented diet
for longer, so when it comes to great apes, the same trick
doesn't always work. And when it comes to us, well …

Baby humans are much the same as baby chimps in that
they try out a hilarious range of movement options before
getting into an adult pattern. An utterly non-random
survey of mothers – whom I know by virtue of being a
member of the same antenatal class – have provided video
and pictorial evidence of human infants going from
immobile to mobile in some truly unexpected ways. There

---

[*] Unless you are a human, in which case it is a nightmare of
precarious things, sharp things, electrical things, hot things, watery
things, poisonous things and other utterly terrifying objects you
used to call 'the totally normal stuff in my house'.

are the rotational experts, who achieve longitudinal motion by means of turning themselves over (and over, and over), rolling towards some distant goal; there are the wounded soldiers, re-enacting baby Dunkirk and dragging their full body weight around with one flailing and gripping arm; there is the classic bum-scooch, which is sitting and, well, scooching; there is a kind of precarious, constant-motion tripod arrangement; there is a sort of serpentine wiggle that might be charitably described as a crawl, but only operates in reverse gear; there is proper tummy-off-the-ground crawling, which, surprisingly very few of the infants in the group opted for; and then there is simply leveraging yourself upright by any means possible, holding on to something (or not) and flinging your feet around until forward motion is achieved.

The interesting thing is that every single one of these remarkable feats is achieved within the first year or so of human life, while a chimpanzee, which is more nimble at birth than a human baby and takes its first steps at six months, goes through gripping, swinging, walking, crawling and all manner of locomotive options before committing to a more adult mode of getting around on all fours at around ten or so years of age. But once human babies learn how to walk, well, that's it. Game over. We are off the breast and on our feet as dedicated bipedalists years before chimps even pick a favourite gait – nearly a decade earlier than they solder up the parts of their skeleton. But we don't speed up our skeletons to match. It's a good decade and a half before we do ours. So, while it might have been important to our primate ancestors, how we move through the world is not the only thing driving the schedule in our skeleton. That means the traces of growth, the pattern of when we solidify our bones into their adult form, is almost totally unmoored from our life history. How did that happen?

Well, not everything about growth is determined by the type of animal you want to become – leggy, brainy, massy. A lot of our pattern of growth is a way of buffering the dangers of an unpredictable world. The environment in which you are doing your growing can step in to encourage – or discourage – growth. Growth happens when you can extract enough energy from the environment to build a baby, and in a sensible animal that means growing more when there's more energy you can extract and going slow when resources are low. We saw in Chapter 6 that many primates are seasonal breeders for just this reason: mama needs to time the most demanding stages of growth – late pregnancy and early infancy – for when she can get the most energy into her own body to pass on to the baby. Stretching out stages of growth is a solid strategy to overcome seasonal food shortages; long pregnancies can buffer some of the seasonal risk (and are consequently harder to reschedule). In the same way, keeping an infant an infant means mama has longer to supplement her baby's growth with her own energy stores through her milk. Keeping the baby on the breast also gives it longer to grow as the sole apple of its mother's eye, funnelling mama's resources directly into its gaping maw and not to any extraneous recipients like siblings. But, for humans, it might just be that we have decided the risk we want to buffer is *later* – in childhood proper.

Growing up is risky and not everyone gets the same chance at growth. As we saw with the run-up to reproduction in Chapter 5, maturity comes to those who eat. What we see in the switching on of the hormonal triggers that launch our ability to reproduce has implications for the growth of our skeleton as well – these are part and parcel of the same complex signalling system. The completion of skeletal growth is, to some degree, regulated by inherited possibilities – this is why the child of two very short people is unlikely to trouble the Los Angeles Lakers' draft picks. But that genetic fate is only a guideline – it sets the parameters for how big you'll get, but doesn't determine whether you'll hit the upper limit or scrape the lower. What determines that is going to be the mix of resources you have to put towards growth, plus any additional signalling for growth to start or stop. There is a great deal of flexibility built in to how we grow for the very important reason that for most of evolutionary history, growing has not been a guarantee. Variation in the availability of food, ecological pressure, competition – these have been our constant companions.

When growth isn't easy, we – and our primate relatives – can employ a few tricks to try to give ourselves the best chance of achieving our full size. One obvious technique is to delay finishing off growth. Waiting a bit longer to switch into reproductive mode and trigger the shift to fusing shut our growing bones gives us the potential to add a few months or possibly even a few years to our bulk. Wild chimpanzees seem to take advantage of this option; a study of a group from Taï National Park in Côte D'Ivoire showed that their skeletons fused in exactly the same pattern as their captive (well-fed and cared-for) compatriots – but sometimes up to several years later. Debates rage over whether the wild chimpanzees grow slower because life is harder or because food is more scarce, but the end result is the same – without being solidly in the black on your energy balance, growth is delayed. Animals have mechanisms for dealing with this – possibly even anticipating it – that we file under the broad heading of 'catch-up growth'. That is exactly what it sounds like and the thing every undersized child wants to hear – that we can still put on a bit of growth as long as the race isn't fully run.

The on/off potential of growth is perhaps one of the most important aspects to consider when we want to understand how we got these long, extended childhoods, with their fast and slow periods of growth. If we think of scaling up the catch-up-growth phenomenon to a whole-species strategy, you can see how a little animal might start to plan when would be the best time to need a lot of energy to burn and when would be the best time to go slow. We need to know the timing of growth to explain the life-history trade-offs – and evolutionary repercussions – of the very unique human pattern of stages of growth, or life-history landmarks. As we've seen in the past few chapters, for us clever monkeys, the mother (and father) have ways to draw out the energetic burden of getting us fed, so our schedules for growing and moving around the world have become unmoored from the normal mammal pattern. We are infants for a shorter stretch than you would expect, but children for longer. The human species is pushing slow life history to its limits.

There are, of course, some basic biological reasons why our 'juvenile period', or childhood, should be different from a lemur's. For one thing, we are a lot bigger, and it takes longer to grow a bigger animal. We also have slightly different proportions to grow – we have worked rather less hard on tails and magnificent ruffs in order to concentrate on brains and, well, actually, some still quite magnificent ruffs; the point being that some of our tissues might have been more costly and taken longer to get the energy in place to build. The environment you're in also has the potential to slow you down or speed you up – highland gorillas stuck with a monotonous diet of leaves plough ahead with growth faster than their lowland cousins who are in constant danger of running out of fruit and so need their mother around longer as a back-up source of nutrition. You might want to stretch out different phases of childhood, or shorten them, according to how risky or demanding they are for the mother (and father) as well. Foraging for two is a lot less demanding when the lump you are feeding is a lump; either a bump in a belly or a relatively immobile infant. Foraging for a frenetic wriggling growing monkey *child*, however, is a whole other level of demanding – and that is as true of little tamarins as it is of little Tarquins.

The other thing we have to remember is that when it comes to growth, the goal isn't just to survive. It's to survive and send your genes successfully down the line. This is what really throws growth for a loop because you can put your money on getting bigger, or you can put your money on making more of you. What you can't do is both, so comparing growth trajectories between ourselves and other primates then needs to take into account the life-history decisions that we've made: big or babies. And just to be complicated, males and females in the same species don't necessarily make the same call. If you remember, we saw that mammals tend to have bigger males than females, if they're going to go down the road to size differences between the sexes.* If you want different sized animals, because, say, you need to become lord of the savannah and its hordes of baboons, or you are thinking of going for a harem, you're going to have to have two different types of growth – a big-male kind and a let's-not-waste-our-energy-female kind. The big question in childhood is when you fit that in.

---

* Unless they are hyenas, in which case the road they have chosen is idiosyncratic, fascinating and has a bite force like being hit by a Mack truck.

The extra size in males is usually due to the addition of a post-puberty growth spurt that occurs in males and not in females. Now this is a curious devotion of resources when we consider the case of modern humans, whose attempts at sexual dimorphism are laughable at best.* It also doesn't make a huge amount of sense for the smaller primates down at the bottom of the family tree, who tend to have either no dimorphism or even to have ever so slightly larger females. For a single-male-dominated social system like the gorilla's, however, traditional evolutionary theories of sexual selection − the asinine ones that insist on manly men and demure little women, and frequently get repackaged as dating advice on the internet − tell us that the male growth spurt is absolutely required to keep in the mating game. But the evolution of man-the-competitor may have been prematurely announced. Given oestrogen functions more or less as the chequered flag at the end of the race to growth, anthropologist Holly Dunsworth has argued we might be reading too much into the size differences between males and females. Those big men might not be trying to impress everyone, peacocking around and beating their chests (to horrendously mix animal metaphors just like I was trying to sell you relationship advice). We might actually be mistaking the reason human males are a teeny bit bigger; perhaps rather than trying to become an ultimate fighting/loving champion, males are just bigger because females are smaller. And females are just smaller because they stop growing earlier so they can put that energy into babies because, as you may have picked up from the theme of this book, babies are important.

---

* Though we certainly seem to be able to identify it in a poorly lit bar at 2 a.m.

However, the strangely extended periods of childhood that primates have evolved to have cannot all be chalked up to sex or size or tissue type, or even environment, and we know this because we can look at the variation within primate species and see that there is still a lot that needs explaining. Chimps and bonobos, whose environments are probably similar to our human ancestors, leave off breastfeeding at around three to four, but hit reproductive age by around ten and might expect their first babies around thirteen or fourteen. This is still awfully fast compared with humans, who (as a gross average) might be off the breast between two and four but don't hit reproductive capacity for another decade – and then don't generally have a baby for another two to four years after that. So, something else is going on with our childhood – and not just the fact that nobody earns enough money these days to settle down.* We must need that time or we wouldn't be taking it. Compared with mammals in general, actually, all primates are keen childhood-extenders. So, what exactly are we taking our time lingering around in childhood for? One explanation is that we are taking that time to learn how to be a better monkey.

---

* Though that's not unrelated – we'll be talking about this later.

## CHAPTER FIFTEEN

# Give a Dog a Bone: How Palaeoanthropology Started to Chase Down Childhood

This old man, he played one,
He played knick-knack on my thumb;
With a knick-knack paddy whack,
Give a dog a bone,
This old man came rolling home.

W e saw in the previous chapter that the real variation in our species versus our great-ape closest relatives comes in the long period between being a baby and having a baby – our slow-grow childhoods and the sort of weird halfway house of adolescence that comes after puberty, but before we properly start reproducing on our own. Our skeletons follow these leads, stitching themselves together at a rate that accommodates our size – whether we are more elephant or more mouse – and the way we want to use our skeletons, whether it is aerial acrobatics or hilarious ground-based side-leaping.* They also take cues from our environment, of course, the availability of energy to turn into height, girth or even new babies, and this can be ramped up or slowed down even on an individual level. But if we want to know where and when and how these choices have been made up and down our ancestral line, we need hard evidence. Like, really hard. Fossil hard.

---

* Sifakas are wonderful and I will not hear otherwise.

The burning question of how we got to be the way we are is why we need fossils.* Fossil remains of our ancestors, immediate and non-, allow us to build up a picture of the choices we have made, and the adaptations we have gained or discarded over time. Something that has changed is very likely to have changed for a reason – usually because it offers some adaptive benefit, but occasionally just because statistics worked out that way† – and that is what reveals both the challenges we faced in our evolution and the solutions we came up with.

Throughout the first part of this book, we've mostly discussed living primates – ourselves and the rest of them, from lorises to gorillas. However, while it's fairly well established that the closest living relatives we have are the chimpanzees and bonobos, we haven't been one big happy family for approximately 4 million years.‡ This leaves considerable room for improvements that have led us to become us and them to become them. This transition may have happened without primatologists to note it in

---

* We really do need fossils. Luckily, most fossils relating to human evolution end up in scientific collections – imagine if we had to do what the dinosaur folks do because so many dinosaur specimens end up in private collections, and go around various rich people's houses going 'hello, could I please come in and do some science, I promise not to get mud on the rug?'

† For instance, the glass-fine, near-translucent hair of the last mammoths might be an example of what is called a 'founder effect', where a genotype that isn't specially selected for becomes more common because of the limited available options in that population – and it is hard to imagine anything more limited than being that last handful of woolly mammoths left alive on isolated Wrangel Island in the Russian Arctic 6,000 years ago.

‡ While the chimps and bonobos have been separated for a mere million years, which is fairly amazing considering how similar they appear (very) and how similarly they behave (honestly not at all).

their stained and dusty field notebooks,* but there is a record of course: the fossil record. With the miracles of stratigraphy and ever so slightly fancier techniques like electron spin dating† we can put the fossils of the primates who occupied this gap between us and the rest of the living primates into a sequence. By arranging the piecemeal clues – some bones here, a few teeth here – on a timeline of millions of years, we are able to work out which parts of us make us *us*.

So, when did we get our human childhood? And why? This is rendered a little confusing by our evolutionary timeline. There are several species of *Homo*, of which we are the only one living – but that didn't used to be the case. The very earliest type of 'us', *Homo erectus*, was pretty clearly different – smaller body, smaller brain and a muzzle-y face only a *Homo erectus* could love.‡ They emerged a bit more than 2 million years ago – but then they stayed around. This is what the latest revolutions in palaeoanthropology have all been about – discovering our ancestors lived in worlds where other types of humans were still wandering the Earth.§ *Homo erectus* leads to a still debated species called *Homo antecessor* dating back to about 1.2 million to 800,000 years ago that gave rise to both

---

* Notebooks not having been invented in the Pleistocene.
† Stratigraphy being the exact equivalent of finding the sweater you wore on Tuesday by digging down through Thursday's shirt, then Wednesday's dress in the laundry basket; ESD being a lot like sticking your shirts in the microwave, under a microscope, and then just getting more complicated from there.
‡ Though we are increasingly unsure of this – the polite way to describe the criss-crossing of genetic lines across species is *introgression*, and we just don't know enough about what *Homo erectus* DNA would have looked like to be able to spot their, ahem, introgressions.
§ And occasionally available for dating.

Lorem ipsum

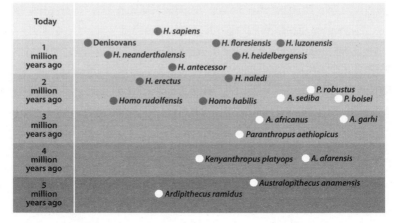

Figure 15.1. *Species through time. H = genus Homo, P = genus Paranthropus and A = genus Australopithecus.* After Humphrey & Stringer (2018).

Neanderthals and, eventually, anatomically modern humans around 300,000 years ago. But in the meantime, *erectus* wandered all over the African and Eurasian continents. We're not even sure when *erectus* disappeared – new dating of fossils at Ngangdong in Java seem to show it still kicking around a mere 120,000 years ago, when modern humans were already well on their way to disturbing the peace and quiet of non-African continents.

During the 2 million years *erectus* was wandering the world, myriad other species popped up – *Homo antecessor* and then those Neanderthals in Europe; *Homo floresiensis*, the hobbit,* in Indonesia; Denisovans in Russia; and now even *Homo naledi* in South Africa. What remains unclear

---

* Apparently species names don't just antagonise palaeoanthropologists. The Tolkein estate is rather unimpressed with calling the little guys from Flores 'Hobbits'. I suggest adopting the neutral term 'Hobbitses'.

is which of these species bred with each other, and when. We know, for instance, we interbred with both Neanderthals and Denisovans, probably multiple times; who they were getting off with in the meantime remains a mystery until we get better ancient DNA coverage. This means drawing a straight line from 'ape like' to 'us like' is impossible. There are evolutionary detours we never expected, in species we have only recently discovered. But making outlandish claims based on the flimsiest flakes of fossils is all part of the fun of palaeoanthropology,[*] so while the paleoanthropologists bare their fangs at each other and argue the details of speciation, we will look at the bigger picture, through the very smallest (but hardiest) pieces of evidence.

Our immediate lineage, *Homo*, are big walkers and thinkers and we can pick them out of a line-up of ape-y fossils by a few targeted areas of growth: an emphasis on big brains (as demonstrated by fossil skulls), longer legs and not much effort put into growing males bigger than females. Before them came the *Australopithecus* types, like Lucy, who came in a variety of different flavours, and combined being generally upstanding members of the primate clade without bothering with too much brain or getting too big for their newly upright britches. They did have some swanky new innovations in the realm of teeth, however. As we go further and further back in time we fall further into the murky Miocene swamp, where some critter lived that was the ancestor of chimps, bonobos and us, about 4 million years ago. At this time, evolutionary decisions were made in building a new type of primate, perhaps starting off with

---

[*] The other fun bit is snarking in the pages of *Nature* at other palaeoanthropologists for doing this.

something like *Ardipithecus* with its strange halfway–up–a–tree-halfway–down morphology.

When everyone still thought of the Enlightenment as a 'good thing', and not a worryingly wobbly step on the ladder of cultural hubris, the fossil record, like everything else, was imagined as a straightforward march of progressive adaptation. There would be a sort of chimp-looking fellow, followed by increasingly human–y chimps, until you got to us. *Et voilà*, evolution! The clearest difference between our ancestral state and our current one, even 150 years ago when people first thought to argue about this, was our brains. Human babies, after all, are very big into the brains. Can't grow enough of them. Compared with monkeys alive today, which is what we imagined ourselves once to have been, our brains have gotten huge and that is a pretty defining human characteristic, *ergo* we should see in the fossil record increasingly brainier chimp-people. This was so achingly obvious to some evolutionary theoreticians that they could not even be bothered to wait for the slow trickle of new fossil specimens to percolate out of the soil, and they were forced to take matters into their own hands to 'prove' that evolution had favoured brains above all else.

In 1912, one Charles Dawson, amateur archaeologist, miraculously produced a remarkable handful of fossil remains from an old quarry called Piltdown in south-eastern England, where geologists had identified a series of gravels thought to belong to the Pleistocene era of about 500,000 years ago. Fragments of both jaws and a skull were painstakingly reconstructed into *Eoanthropus dawsoni*, and 'Piltdown Man' was revealed to the public as a remarkable 'missing link' – ape jawed, but capacious of cranium. It took 40 years for enough access to be granted to the specimen to be dated, when it turned out to be one tenth the age advertised, and then it took almost exactly 100 years more for Natural History Museum of London science to be turned full bore on this most questionable of ancestors. Being the person in possession of the 'big' computer in Anthropology used for viewing CT scans outside of the dedicated lab downstairs at the NHM, I had ringside seats as a team of my colleagues slowly unpicked Dawson's 'dawn-man'. The teeth, it turns out, weren't human or even human ancestor – they were orangutan, and the CT clearly showed where they'd gotten gravel stuck in with the putty as they were jammed into Piltdown Man's jaw. And that head? Bits of medieval people.*

The saga of Piltdown Man may end up as an entertaining footnote in the annals of human evolution, but among the Frankensteined pieces of the museum collection there lurks a more severe error we would be wise to keep in mind. This is the error of conceiving the human story as a narrative of descent, like you would expect in a pedigree dog or a British royal. We *cannot* tell the story of us like they do in Genesis, with a line of 'begats' that stretches 4 million years in the past. It gets all mixed up, with some groups breeding back and forth so no one can quite tell which grandfather is which, and some dying out completely. We can talk about animals that walked this Earth before us, and the features they shared with the

---

* As one of the premier research collections in the world, the NHM has a lot of bits of medieval people collected during archaeological excavations – and curation in the bad old days was not as advanced as it is now. The scope of these advances can be easily judged by the amount of time it takes between mentioning a previous curator's name to a new curator and the eruption of violent oathing.

animals that came before them and those that came after, and work up ideas about which traits were 'passed down' – but we cannot get the resolution (yet) to detail all of these things with absolute certainty. Sometimes traits that look human show up in one species, only to disappear in the next – and vice versa. Bigger brains seem like a clear difference between us and our last common ancestor, and there is definitely more going on upstairs in our species than in whatever Miocene tree-leaver we descend from. But what Dawson got wrong, and as the caves and canyons of Africa are slowly revealing, was *which* bit of us kicked off the changes that took us from furry vegans to four-star vegans. If only he had stuck a more modern-looking jaw on an ape-y cranium like the real-life fossils palaeontologists have actually found, maybe we wouldn't have noticed.*

The real story of how we got to be the animals we are is out there, however. While rare, the remains of growing bodies are preserved in the hominid fossil record, giving us a huge depth of time to consider the path we have taken to get from the ape-like common ancestor we share with chimps, to the place we are now. Working out how this affected our childhoods, however, is rendered immensely more complicated by the fact that we have so few skeletons that have gone into the mud during the period that they were still growing. Once bones are done growing, like the bones of an adult, there's no (good) way to guess how long it took them to grow. Think of the most famous fossil there is, for instance: Lucy, the *Australopithecus afarensis*. There is no way to tell exactly how long it took her – one of the earliest upright

---

* We would have noticed. If you ever want to be bathed in an absolutely unholy amount of descriptive detail, read a palaeontological species description.

walkers in our family shrub* – to grow her femur from 20cm to 28cm; where she would have sat on the elephant-to-mouse scale of speed of skeletal growth; and what using her legs in the manner we think she did would have done to that schedule. All we *do* know is that she had a fully formed femur when she died, so we can say she was an adult. Similarly, the fragments of her hips show that the six childhood bones of the pelvis (and dozens of bony end pieces) have coalesced and come together into the two halves of an adult. But what can we say about the stages of growth, if we only have the end result? How do we know if Lucy lived fast or slow? How do we find out if she had a childhood like ours?

What anthropologists need to be able to do is to compare two, say, upper leg bones found at some million-year-old fossil site and tell you whether they are different stages of growth – or different species. The former, I am pleased to report, is one of our greatest skills, and it is the source of the evidence we need to understand how humans got our childhood.† So what *is* our evidence base? How do you calculate childhood from what's been left behind hundreds, thousands or even millions of years later? Some of the same information from our skeletons that we saw mapping out the differences between ourselves and other primates in the previous chapter can be deployed to figure out how our ancestors grew. Of course, what we really need to understand the phenomenon of human growth is a comprehensive sample of the young of all of our ancestors, culled at different ages and stages of growth, but that's not what we have. What we have is a semi-random speckling of tantalising fragments and chance finds – and that has to be enough.

---

* The human evolutionary tree is so convoluted, knotted, and debated that it's arguably more accurate to imagine the tangle as a particularly unrestrained box hedge.
† The latter is bloody difficult.

Nariokotome boy (or KNM-WT 15000, to give the proper scientific name)[*] is perhaps the best-known juvenile to have emerged out of the fossil record, and he caused something like two decades of ferocious arguments because his skeleton has hallmarks of being either very ape-like in its growth pattern or, conversely, very human. The difference is significant – chimps reach skeletal maturity almost twice as fast as we do, despite being about the same size, so his whole childhood hangs in the balance. With a relatively large skull and a pelvis and limbs arranged for upright walking, the 1.5-million-year-old fossil find was identified as a juvenile *Homo erectus* – the very start of the hominid lineage, falling this side of the dividing line between big and small brains that marks out our earlier ancestors, like Lucy's australopithecines, from the hominids. Nariokotome's skeleton is so important because it is unfinished; the bones of his skeleton were still in a relatively unfused state.[†] The question then becomes – did Nariokotome boy grow more like a chimp or a human? Fast or slow? Because understanding that lets us trace the path we took to get to us.

Nariokotome was still in the process of growing when he died. His elbow, for instance, was only partially complete. The small end caps that make up the hinge joint of the elbow come in three unsophisticatedly shaped parts, two of which first fuse together to make the bearing for the

---

[*] This fossil is also known as 'Turkana Boy', presumably to keep others from making the mistake I once made and confusing the pronunciation of the proper name with a certain fictional office tower from the cinematic masterpiece *Die Hard*.

[†] Well, we call him a 'he', but this is a little tenuous because while his pelvis has some of the characteristics of a male, we don't know for certain what shape his pelvis would have eventually grown to be.

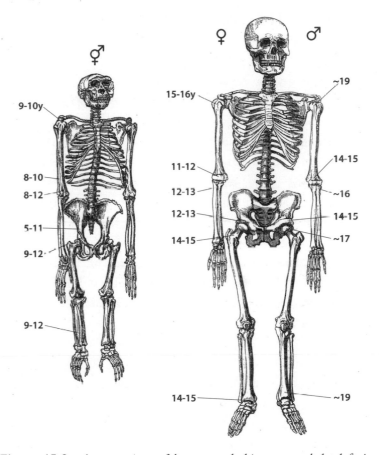

Figure 15.2. *A comparison of human and chimpanzee skeletal fusion patterns. After Brimacombe (2017).*

bones of the lower arm to rotate around before glomming on to the main body of the humerus. For Nariokotome, those two parts had all adhered together and started to fuse to the upper arm, but there was still a clear line where the process was ongoing, and the third and final part had yet to attach. For a human boy, that would start at about 12 and a half, and finish by about 15 years old – and a little earlier if he was a she. For a chimpanzee? The same process would take place between about seven and a half years old until about ten and a half, as you can see in Figure 15.2 above.

Judging from the level of 'doneness' of his skeleton, he (if it was a 'he') either died after 10–13 years of a human-paced childhood *or* 7–8 of a chimp-paced one.

The problem is, it's incredibly difficult to make out a pattern from just one fossil. There have been other juvenile fossils that are slowly helping to fill in the several-million-year gaps between ourselves and our closest living relatives, but not that many – turning into a fossil is a very rare achievement.[*] This is why finding fossil kids is such a major event and how one little body can rewrite our whole understanding of human evolution. There are only a handful of juvenile bodies known. There are remains of an adolescent *Homo erectus* from Dmanisi cave in Georgia that are at a slightly later stage of maturation than Nariokotome, adding perhaps two years to either the human-like age or the chimpanzee-like one.

There is more intriguing information available from the older australopithecines, with skeletal information available from juvenile *Australopithecus sediba* and even an infant *Australopithecus afarensis* – the same species as the famous 'Lucy'. The juvenile *sediba* has a humerus in a state of fusion not unlike Nariokotome boy's, but having gone one stage farther to start fusing that final piece of the three-piece elbow end cap. With *sediba*, we know the adults look pretty ape-y; so we might suspect a much more ape-like pattern of growth overall and estimate that it is much younger than the 13 odd years a human with the same skeletal fusion would be – somewhere between 9 and 11.

Most recently, there is the astonishing find of an entire cave full of a new species of small hominid – *Homo naledi*,

---

[*] Mudslides help. Also, caves.

whose bits and pieces are scattered on the floor of the
Rising Star cave in South Africa, nearly a mile from the
entrance, deep in the cave system.* The finds from the
Rising Star excavation are still being processed – indeed
more are still being uncovered – but there seem to be a
good number of individuals represented and, on April
Fool's Day 2020, a brand new 250,000–300,000-year-old
juvenile was presented to the world.† Now, *naledi* is an
odd duck, an utterly unexpected‡ small-brained, small-
bodied upright walker with a chest like an
australopithecine, but potentially the brain structure of a
much more contemplative beast. It is interesting to
wonder what exactly our ancestors might have thought,
encountering this diminutive cousin as we moved into
South Africa some 200,000 years ago – unless, of course,
the timing of the *naledi* extinction turns out to be a bit
like what happened to all those nice Hobbitses on Flores
or those tasty mammoths – and something we really wish
we hadn't asked questions about.§

Specimen DH7 from the Rising Star cave is made up of
60-odd fragments of juvenile bone, some of which clearly
fit together in the lab like so many hundred-thousand-
years-lost puzzle pieces, revealing the broken bits of a

---

* And one 11-inch gap away, which is why excavation is limited to
the very slight, and those willing/able to voluntarily dislocate their
shoulders, like the hobbyist spelunkers who discovered the cave.
Ugh.
† Apparently an amazing discovery and not an April Fool's joke.
‡ Which is something of a trope in palaeoanthropology, thank you
Denisova cave.
§ There are also questions to be answered about the demise of *Homo
floresiensis* (the 'hobbit' fossils) around 50,000 years ago – the same
time *Homo sapiens* swept into the neighbourhood.

juvenile who died at nearly the same skeletal state of fusion as the *sediba* adolescent and Nariokotome boy. This lets us ask the million-dollar question: how old? *Homo naledi* lived at the same time as the later, brainy, big hominids like Neanderthals and ourselves, and has a bigger brain than *sediba* – perhaps they too had a slower growth track? In which case the same set of skeletal features in *naledi* that we see found in *sediba* would put them closer to 11 to 15 years, rather than 9 to 11.

We simply don't have the evidence base to come down conclusively on the speed of growth far back in our ancestral line (and the bits we're not sure are even ancestral) with the evidence from fossil skeletons. So, what about trying to work out something a bit closer to home? Neanderthals are the second most plentiful *Homo* species[*] in the fossil record, after ourselves, and we now know that they were at least close enough relations to breed with. Far more so than *Homo naledi*, they look like us – in size and, importantly, in the proportion of brain to body and associated with archaeological finds that suggests they may have shared a lot of our cultural abilities too. They are also closest to us in terms of time, if not space – Neanderthals are a Eurasian phenomenon, and not linked in to the major thrusts of *Homo sapiens* evolution occurring in Africa in the past 300,000 years. And because there are so many of them in the fossil record, we know a lot more about them. Sometimes you just have to have the numbers, and you get lucky and get an entirely new piece of information.

---

[*] There is a lot to say about even the idea of 'species' in hominids, but no one put it better than Inigo Montoya: 'I do not think it means what you think it means.'

Take, for example, the case of 'J1'. The cave site of El
Sidrón, in Asturias, north-western Spain, has yielded a
so-called 'tunnel of bones' – the mixed pieces of several
Neanderthals deposited in potentially grisly circumstances
around 48,000 years ago.* Analysis of ancient DNA tells
us that specimen J1 is the probable son of an adult female
whose remains were also found in the cave, and the
sibling of a nearby infant. The skeleton is well preserved
and, most importantly, has a very nearly intact skull – so
much so that we can take a good guess at how big the
brain was compared to how finished the skeleton was. If
it had been a modern human, J1 would be about six to
ten years old based on how much of the skeleton had
soldered itself together. However, given how big
Neanderthal brains get, researchers estimated that J1 had
only completed about 87.5 per cent of its total brain
growth by that stage – whereas a human child at the
same age would have already hit 95 per cent of future
brain capacity. Our whole skeletal schedule is thrown for
a loop: if J1 is less grown at a later age, could Neanderthals
have possibly grown *slower* than us, reigning champions
of the delayed maturation and true holders of the title of
longest childhoods? Luckily, there is a better way to
answer the question of how we got the childhoods we
got – and it's staring us right in the face.

---

* The bones of El Sidrón show lots of cuts and damage that suggest
they were butchered and eaten, though by whom, it is hard to say.
The happiest possible interpretation is that Neanderthals had some
really intense funeral practices, akin to those recorded for some
human societies – but the chances of a whole kin group needing a
joint cannibal funeral are somewhat slim.

## CHAPTER SIXTEEN

# Grandmother, What Big Teeth You Have: How Teeth Gave the Game Away

Grandmother, what big teeth you have!
All the better to eat you with, my dear.

The Taung Child has been dead for almost 3 million years, and yet remains one of the most important discoveries in palaeoanthropology. The story of how it came to be found – and recognised – carries the hallmarks of early palaeontological finds, which is to say a degree of whimsy that is no longer considered appropriate to systematic scientific endeavour. In the 1920s, Raymond Dart, the not-yet-famous palaeontologist, was teaching at the University of Witwatersrand in South Africa when he hit upon an ingenious scheme of offering his students a £5 cash prize for bringing in the most intriguing anatomical specimens for his collection, therefore guaranteeing students would spend a portion of their holidays keeping an eye out for bones, fossils and other treasures that could be used in his teaching.

Enter 22-year-old Josephine Salmons, Dart's first female student, who was working as an anatomy demonstrator in 1924 when the holidays fell and she went off to visit some family friends, the Izods, who ran the Buxton Limeworks in the small town of Taung. Spotting a baboon-like skull on the mantelpiece that the family said had come out of the rubble from blasting open the mines, she begged the loan of it and presented it to Dart once she was back at university. Dart recognised it as a fossil baboon, rather than a modern one, and set off immediately to see if there was anything more left at the Taung mines. A worker there at the limeworks called De Bruyn had been building a collection of curiosities he'd found, and eventually two massive crates of fossils were shipped to Dart's house, arriving in the drive as he was supposed to be getting ready for a wedding reception. Barely restraining himself long enough to do his duties as best man, Dart threw himself into work on the material. He spent the winter using his wife's knitting needles to slowly chip out the fossils, and was finally rewarded with a near-perfect endocast of a primitive fossil brain and the better part of a face that had never been seen before: that of *Australopithecus afarensis*. The discovery was written up as the missing link between ape and man, and Dart became a palaeontological superstar.[*]

No bigger than the size of two closed fists, the head of Taung Child is decidedly small, with a brain even a modern chimpanzee would be loath to write home about. But what caught the world's interest were the human-like features that Dart described – in the teeth and lower jaw, and the central location where the spinal cord goes into the skull so that the skull sits up high when its owner is walking on two legs (instead of towards the back of the skull, so the skull sits forward like a dog or an ape). This is the very thing that fake-news fossil Piltdown Man failed to take into account. Now, not only was the Taung fossil a shockingly small-brained bipedalist, it was a child. This

---

[*] Josephine was written out of the story, as was De Bruyn. And now you know how palaeoanthropology works.

offered an amazing new opportunity to ask entirely new questions: was Taung the origin of our human story? How much of a child was it, how old would it have been? Did it grow like an ape, or like a human?* Taung was clearly too small to be comparable to the skeletons of modern humans, and didn't have enough of its skeleton left to match up the schedule of bone fusion. Dart instead turned the best evidence he had to hand – and the one thing that can tell us for certain how long it took our fossil ancestors to grow: teeth.

You may have seen this coming. You are probably already aware that teeth appear, disappear, then appear again in our mouths while we are growing.† English speakers of a certain vintage will recall the teeth-clenchingly saccharine holiday ditty 'All I Want for Christmas is My Two Front Teeth'. They may even recall that it was traditionally performed, or at least taught to, kindergartners or reception-year pupils. This is less to do with the steady erosion of decent music from the holiday repertoire and more to do with the fact that it is amusing to make a room full of five-year-olds sing about not having their two front teeth because *many of them do not have their two front teeth.*

A growing child is essentially no better than a shark, with a mouth containing both the diminutive dentition of childhood – the primary, or milk teeth – and up inside their jaws, the growing fangs of adulthood. Studies of

---

* The technical answer to this is 'neither' – the Taung Child never completed growth. Three puncture wounds, spread from the cranium and into the eye orbits like the finger holes in a bowling ball, give a potential reason why: there is nothing an African Crowned Eagle likes more than a tasty small primate. Whether this is an excuse for the lingering agoraphobia in our species, however, is best left to psychologists.
† Those teeth can disappear again too, but that's on your own head. Floss.

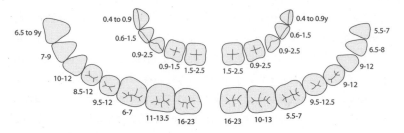

Figure 16.1. *Age that teeth erupt into the human mouth in years. Upper teeth are to the left of the image, lower teeth to the right. The more numerous adult teeth form the lower row below the baby teeth at top. Data from Liversidge (2003) and AlQahtani et al. (2010).*

living children from all over the world have been carried out to establish that the pattern of teeth erupting into the mouth, then falling out and then being replaced by their adult counterparts, is incredibly standardised – especially when compared to the growth of the skeleton, which as we learned previously, can be compromised by nutrition and other factors. While there is variation in the timing of teeth, we can usually confidently look into the mouth of a growing human (or a chimp, or a lemur) and pick their age within a few months to a year or so.

This is because teeth emerge into the mouth when we need them. Perhaps one of the most insightful questions I've ever been asked in a classroom – and the one that left me the most bemused, because it was such a simple thing – was the question of *why* we have two sets of teeth. Fully prepared to hold forth on the subject of how to identify a right upper second molar from a left upper second molar, I was rendered speechless because, on the one hand, the answer is very simple – when we are small, our mouths are small, and we need small teeth to fit. When our mouths are big, the small teeth are too small, so we get rid of them for a bigger set. This is a sufficiently noticeable phenomenon that even Aristotle got in on it, though he managed to get it rather fantastically wrong.* On the other hand, there are a host of evolutionary pathways that we could have gone down and did not, ranging from the singular tooth-horn of that nautical unicorn the narwhal, to the not-quite-teeth millstone grinding plates of the sturgeon;[†] so the question is immensely complicated by all of the things that have happened in the roughly 500 million years since teeth were invented.

However, why our teeth appear in our mouth at the times they do, looking like they do, is intricately tied up in what we need our teeth *to do*. Teeth are inexorably involved in one of the most important businesses of living: eating. Like the beaks of Darwin's finches, they are specialised for the food we eat. Fish crack crabs with great lumps of bony plate while cows are possessed of a handful of grass-snipping teeth and many long-ridged teeth perfect for grinding their fibrous diet. Primates, on the other hand, have the Swiss Army Knife of dentitions. A mix of tooth forms allows us to exploit a variety of foods – or at least, has done, in various iterations of our ancestors.

Incisors, the teeth at the front of your mouth that are frequently lamented by singing schoolchildren, are for biting – incising. Canines, the dog teeth (or vampire fangs), are for puncturing and tearing, and, if you are a baboon, impressing the ladies. Premolars are a halfway house between the gripping canines and grinding ridged teeth at

---

\* Camels do have upper teeth, women don't get wisdom teeth in at 80 and the man seems to have been absolutely baffled by tusks (see Book II, *Historia Animalium*).

[†] Narwhals essentially have one tooth that grows extraordinarily long, forming the (unfortunately, for them) highly collectible narwhal spear. Sturgeons technically don't have teeth at all, but do have eggs that taste delicious and sell for a small fortune, which again, is unfortunate for them.

the back, the true molars. Most adult primates have the same set number of these – for each quarter (upper, lower, left, right) you get two incisors, one canine, two premolars[*] and three molars.[†] Young primates get a slightly reduced offering with the same front teeth, but no premolars and just two molars for chewing. While some species have seen fit to alter their teeth to suit particular needs – the lemurs, for instance, turned all their front teeth into a comb for their luxuriant fluffy coats – most primates have the same set. And what is very interesting when we think about how we grow is that these teeth appear at roughly the *same point* in individual primate's lives. The arrival of teeth into the jaw is the start point of a developmental clock that is visible without having to resort to X-rays and CT scans, making it one of the most important starting points for working out the age – and development – of an animal. And when we combine this with the state of development of the child – which teeth have erupted, which bones have got how long and which have fused or still have room to grow – we can reconstruct the evolution of our own unique pattern of childhood.

Teeth are clearly a requirement for eating,[‡] but we do not get them all at once. You also tend to have more or less teeth at birth depending on the size of lemur, or monkey,

---

[*] We evolved from things that used to have four premolars. This would be neither here nor there except the premolars we kept were technically the third and fourth in the sequence, which means anthropologists, and only anthropologists, continue to refer to your two premolars as the third and fourth in memory of their long-lost fellows. This wreaks havoc on undergraduate exam-takers.

[†] The last one is your wisdom tooth and, actually, you might not have it at all if you're young – or lucky.

[‡] For most vertebrates, that is, and certainly for us prior to the invention of the blender.

or ape you are trying to grow. The smaller animals far down the family tree, like lemurs and lorises, are born with some teeth already erupted, ready for a rapid transition to adult eating – the lizard-crunching tarsier is ahead of them all with some 20 teeth already in place by birth. Some of the smaller monkeys also have teeth ready by birth, but by the time you get to the bigger animals like great apes you are presented with a baby that is toothless for months. When these missing teeth do arrive, they are not actually our 'real' teeth; all mammals start off with the much smaller, reduced set of teeth that fits into our baby jaws, before gradually replacing and supplementing these more delicate baby teeth with the adult set.*

Generally, there is a staged deployment that makes the most of the available jaw space and the immediate needs of the animal. We are not obligate grazers like a baby foal, for instance, so we do not immediately need big bumpy check teeth for mashing grasses into nutritive submission. We began with teeth designed for not much more than nibbling, and only when the milk begins to run out – or be withheld – do we need the calorie-crunching heft of big molar teeth. The eruption of our *adult*-sized first molar† often marks just such a momentous occasion, when we finally kick mother's milk. The eruption of this first grown-up tooth is a pace set across an entire species by evolution, at least in large part, rather than by the needs or

---

* A very small number of mammals cycle through teeth rather than having a baby set *per se*, but no one ever thought manatees, kangaroos or elephants were normal anyway.
† Follow along at home: this is the first wide, bumpy tooth after you run out of thinner, pointier ones when you run your tongue over your teeth from the front to the back of your mouth. You should have three in a row, unless you haven't got your wisdom teeth (those would be the last ones in, at the back).

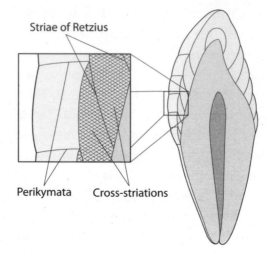

Figure 16.2. *Section of a tooth showing regular growth structures: the striae of Retzius and periykmata form every 6–12 days, and the cross-striations every day. After Moggi-Cecchi (2001).*

actions of individual mothers or infants, and the tie between the end of breastfeeding and the emergence of our very first adult molar gives us a critical insight into past growth and development. In many primates, there is a very strong relationship between the emergence of that first permanent tooth and the end of breastfeeding. This tie holds across several living primate species, giving us very good reason to believe that the first-molar-breastfeeding rule should hold across fossil ancestors and other attenuated branches of our family tree as well.

But, of the great apes, only gorillas seem to follow the pattern predicted by the rest of our clade. They, like chimps and orangutans, get their professional chewing teeth in at around three to four years of age and they stop breastfeeding around the same time. The chimps and orangs keep breastfeeding for a few more years, while we go completely the other way. By the time we get our first adult molar between the age of five and seven, we've already stopped breastfeeding, probably for several years. That's not just some weird modern thing, either – as we saw previously, very few societies breastfeed for longer than a few years.

Louise Humphrey of the Natural History Museum in London, who is directly responsible for many good bits of my career and not at all to be blamed for the rest of it, has a distinguished record on investigating exactly these questions* and suggests that one key factor in understanding the human switch to shorter breastfeeding times is our need to grow those enormous baby brains. For the vast majority of primates, almost all of the eventual total brain size is achieved by the same time the first adult molar starts to work – the two are tied together in developmental sequence. In humans, however, we are nowhere near finished with our brains by the time we get our first molar. But as we trek through brainier and less brainy iterations of hominids, we should be able to track when the major milestone of childhood independence has been achieved.

Taung Child has their first adult molar in place, matching a six-year-old human child's. A quick X-ray allows us to see the rest of the teeth, developing inside Taung's jaws. We can see which teeth have finished all the way to the roots and which are still growing. Comparing the schedule of when modern human teeth emerge into the mouth with the array of teeth seen in Taung's tiny jaw, Dart concluded that if Taung was human they would be six years old. But what if the pattern of dental development was more ape-like

---

* And has the rather more dubious honour of infusing me with an unshakeable interest in these questions by virtue of employing me in my first ever postdoctoral research position; she is expected to make a full recovery.

and the molar came in earlier? This is precisely the question that needed to be answered and in the late 1980s a flurry of research was directed at it. Arguments bounced back and forth over whether the stages of the teeth matched an ape-like pattern or a human-like pattern better, anthropologists writing green-ink letters about each other's theories in the pages of *Nature*. If Taung had a human pattern, this would mean that, before we even reached the stage of increasing our brain size, our ancestors were already on the way to that long slow human childhood.

There is one other source of evidence, however, that could tell us if Taung was headed for a human – or an ape-like childhood. That is the internal clock locked into the tiny microstructures of their teeth – a fossilised record of daily growth. If the daily march of growth inside Taung's tooth was the same as that for modern humans, it would be clear that Taung had reached the developmental stage it had – molar out – in the same amount of time a human child would have. If, however, that same tooth had taken less time to grow, then it would mean that Taung had hit that developmental stage *faster* than a human child. I have had the immense privilege in my career to work with Christopher Dean, who has been intimately involved in resolving the raging debates around the timing of childhood in our ancestors. With colleagues, he has gone through and painstakingly counted both the growth lines inside teeth and without in a number of fossil hominid specimens.

In the case of australopithecines, the group that Taung belongs to, there are *more* lines on the teeth – and therefore more weeks of growth – than on an even earlier species, *Paranthropus*. So Taung and kind grew more slowly than earlier species – but did they grow the way humans do? The answer is an emphatic 'no'. Judging by the number of weekly ridge-lines on the teeth, Taung grew nearly twice as fast as modern humans. That means that Taung wasn't six years old when they died – they were about three, and headed towards an accelerated ape-like childhood.

Now what about our other mystery hominid child, Nariokotome? His teeth are almost all adult – he has lost all but his baby fang teeth, and the adult ones have emerged into their final useful positions. We actually know that they were in a useful position because when teeth are in use, they get worn down; either from smacking into something equally hard, like their opposite number, or from the fibres and grit in food, a phenomenon dental anthropologists call 'wear'. The surface of the enamel becomes abraded, at first microscopically, then you start to lose whole swathes of surface and the teeth do the job they were meant to do. The longer the tooth is in use – particularly in an era before we had foods like jelly – the more worn down it gets, until eventually the whole chewing part is gone and the tooth eventually collapses and falls out.[*] Nariokotome probably hadn't had his adult teeth in place too long because they had only a small amount of wear on them.

The X-ray's all-penetrating eye reveals that those adult teeth aren't even actually completely finished growing. The long roots that hold the teeth into the jaw haven't quite extended their full length and their tips are still open, a sure sign in teeth that they are still a work in progress. The pattern of development – mostly grown – in

---

[*] This is by design – and why you have three molars that emerge at different times, rather than one big one that comes out once. It is also the origin of the advice to never look a gift horse in the mouth – horses wear their teeth down something fierce with all that grass, but when they run out of molars, well, they run out the clock. So, checking a horse's teeth tells you its age and, by proxy, the value of the gift you should just be graciously accepting.

Nariokotome's mouth is the same that you would find in an average 10-year-old human child's. That is more than two years younger than his skeletal age predicts – something that falls well outside the statistical bounds of modern human experience. Turning to the exacting record of lines on teeth, Christopher Dean and colleague B. Holly Smith reckoned Nariokotome had actually reached about eight years of age at the time of death, meaning he grew much faster than modern children. This means the life history modern humans enjoy – what Dean and Smith have called 'live fast, die old', with 'live fast, die young' available on occasion – wasn't yet part of our evolutionary strategy when our first *Homo* ancestor began to spread across the globe. The long, slow process of a drawn-out puberty and adolescent growth spurt that results in our modern-human delayed adulthood is much more recent in our evolutionary history.

What of all of the hominids who came down the line after *Homo erectus*? After all, Nariokotome was only the beginning of the lineage, and there are now more than a handful of species we know to have emerged since then with varying similarities to our modern selves. Some, like *Homo naledi* and the hobbit, do not appear to be very closely related to us large, fat-headed, flat-faced apes. They are small-bodied, with mixes of hominid features that drive palaeoanthropologists crazy, but suggest a very different evolutionary path from the one we have taken. There is, however, plenty of compelling evidence for big-brained beasties more like us, the first and foremost of which has to be our rather-more-than-kissing-cousins the Neanderthals. Neanderthals spread from the Levant all the way into northern Europe starting sometime around 350,000 years ago and stretching right up until near-recent history, 40,000 years ago. They were ever so slightly shorter than us, but thick with it, all barrel chests and massive skulls. Their brains were, statistically, bigger than ours, and despite centuries of 'cave-man' stereotyping, can't have been that off-putting – there is clear evidence of Neanderthal DNA lurking in the genetics of most modern people from populations outside of Africa. Even better, there is a wealth of Neanderthal fossil material that allows us to ask the exact same questions about growth that we have for Taung and Nariokotome.

Traditionally, Neanderthals were assumed to have a human pattern of development, given their similarities to us in size and particularly in achieving such big brains, the factor presumed to be so important in dictating our own pattern of development. But when it comes to their teeth, a count of the near-weekly chronological lines shows they grow faster than the average modern human teeth – suggesting they had more accelerated childhoods. Neanderthal teeth are on the speedier end of things, but not entirely out of the range of modern human potential, so it is hard to know whether the few teeth we have found are extremes, or if they represent a normal speed for Neanderthals that was a little bit faster than our own. Looking at teeth from the much earlier Sima de los Huesos material suggests that earlier *Homo* species were growing their teeth faster than we do.

Fascinating new evidence from China, dating back to sometime between 100,000–200,000 years ago – roughly the same age as Jebel Irhoud, half the world away – has possibly even brought us our first ever look inside the teeth of a new contender in the story-of-us saga: a *possible* Denisovan. Denisovans are my favourite recent non-human hominid for the sheer unexpectedness of them; the chance discovery of their utterly different DNA from a tiny finger bone had the exact same effect as the Spanish Inquisition in the famous Monty Python sketch – no one was expecting it. These mysterious hominids interbred with Neanderthals and with some archaic versions of us, further complicating the neat branch-and-tree diagram of human evolution. Just to add to the confusion, there is not enough Denisovan skeleton around to be absolutely sure of what shape they were, meaning that we could have already found skulls or other bones of theirs and simply not known.

This handful of kid teeth from the site of Xujiayao is a truly thrilling discovery, because it means we might finally be getting more hard proof of these most abstract of semi-ancestors. But because we *also* don't know quite what the DNA is from older species like *antecessor* or *erectus*, it's quite hard to come to a solid conclusion on whose teeth these really are. They could belong to one of the earlier types of *Homo*, or some kind of thing we've written off as Neanderthal, or possibly even a very early wandering anatomically modern human, now that we know our species stretches back to at least 300,000 years. But when they were popped in the synchrotron by Song Xing, in a method pioneered by dental anthropologist Tanya Smith and beamline scientist Paul Tafforeau, they were shown to be well within the bounds of modern human variation in terms of how fast they grow. The laser doesn't lie: whatever was growing in Xujiayao 100,000–200,000 years ago was growing like us.

The million-dollar question here is whether the pattern of growth we see in our species today is entirely our own invention – and it looks as though the answer might generally be 'yes'. When we look back in time at skeletons that look like us, anatomically modern *Homo sapiens*, we see exactly the same pattern of growth we have now. Counting the lines on the outsides of the teeth of a child that died at the site of Jebel Irhoud in Morocco almost 160,000 years ago showed that the child had about seven years of tooth growth under their belt when they died – and, critically, the amount of tooth they had grown in those seven years was well within the range of modern human variation. The speed of dental development tells us that the rest of the hominid line seem to have had just a little less time to spend on childhoods, but a lot remains to be seen;  maybe we're just finding really fast Neanderthals. After all, we have gone in only a few decades time from seeing humans as the arrowhead describing the arc of evolutionary progress, flying far above the fray, to realising we have had a much more meandering slog through a world full of species that were so much like us that we bred with them.

Those other species, however, are gone. Traces of other radiations of the hominid clan remain in some of our genes, but unlike the rest of them we are still here. And we have to wonder why that might be. Over the last million years we can see, locked into the hard white enamel of shiny fossil teeth, the emergence of a radical new form of living: one that goes slow through childhood, fast through reproduction and carries on long after the baby-having band has gone home. While there is a great deal of caution to be argued for, especially in trying to compare a handful of developmental clocks from across millions of years of evolutionary time when we're only now starting to understand the variation in our own species, there is a pretty clear path towards a long slow childhood that wends its way to us down the millions of years. This is our unique *sapiens* pattern, the one we have derived from all of the possibilities offered by our primate relatives and then forced into new directions that suited us best. We can confidently say that we evolved to be children for longer, which rather begs the next question we will turn to in this book: what on earth did we do that for?

## CHAPTER SEVENTEEN

# The More We Get Together: the Importance of Social Learning

The more we get together,
Together, together,
The more we get together,
The happier we'll be.

In our lineage we grow slow. Other animals grow slow because they can take the time to invest and so they grow expensive babies. We also grow slow because we build the most expensive babies we can possibly build and one way to buffer the risk of throwing all your energy into baby growing is to drag it out a little. We have these long childhoods so we can invest more and more into our offspring, making them into the best, brainiest babies they can be. But we're not just building the embodied capital we saw in the first chapter and investing in the physical growth of our babies. We spend childhood training our babies up to be human and we grow slowly because we need that time – time to learn the rules of the road and time to learn how to be a better monkey.

The animals in our lineage that took their time did better than those that didn't – and that got us *us*. In the first part of this book, we saw the initial phases of growing a human – conception up to infancy – were really about investing in physical capital. Parents are embodying capital – taking their own resources and literally feeding the cold hard nutrient cash that they could be using themselves to their children, whether it's filtered through a placenta or at a remove by burning up calories carrying tiny baby butts hither and yon. Childhood doesn't put the kibosh on investing in physical capital for your children – *au contraire*, you have to feed them even more now. But as a baby takes its first steps towards independence, it is also taking its first steps towards an entirely new kind of 'money' sink.* Childhood is the period when you can start to invest social capital. Social capital is what you have when you have friends and networks of support. And social capital is what you can leverage to make sure that your kids have the friends and supporters they need, because – for primates at least – somebody has to *teach* you how to be that better monkey.

All primates – and many other animals as well – do some amount of learning from the caregivers and/or other members of their groups. Social learning emerges through interaction with others and is a tool for consolidating skills and behaviours that can't be hardwired and are too important (or too complex) to leave up to random experience to impart. The urge to avoid drowning, for instance, is not a learned behaviour, but a single monkey-falling-in-a-pool experience should be enough to 'teach' that particular monkey about the advisedness of entering the water. Social learning, by contrast, is when that monkey learns, from caregivers and other group members, that you have to be a bit careful about it, but hanging out in a pool can be a rather nice thing to do.

This is exactly what happened in the case of the famous snow monkeys – Japanese macaques – that inhabit the hot-spring-riddled regions of the northern Japanese island of Hokkaido. Nearly everyone is familiar with the iconic images of these tawny-haired primates relaxing in a

---

* Literally. Apparently it's totally normal to pay £50 for baby shoes. That's £50. For shoes. For babies.

steaming hot spring, sometimes long enough to acquire photogenic clumps of snow as headgear.* This is hardly a long-term evolutionary behaviour, however. Displaced by ski resort development in the 1950s, a troop of macaques were coaxed to hang around the hotels being built around local hot springs by an animal-loving railway employee and an enterprising hotelier. These enterprising souls began encouraging the monkeys – apples were the key, it seems – to hang around, and by 1962, the first brave young monkey was observed to enter the (until then, entirely human-oriented) hot tub at the Korakukan Hotel. Over the following years, the monkey's friends and female relatives had all joined in. What clinches the activity as an episode of social learning – rather than just a case of all the local monkeys at some point or another falling into the hot tub and deciding they liked it – was that over time as new males joined the group (females stay together in this particular species, while males are rovers) they did not necessarily 'know' about the hot tub – and weren't seen to take a dip.

Social learning plays a critical role in transmitting the information that primates need to stay alive down the generations. While learning to enjoy a bath might not seem critical, it is a behaviour adopted in the same way – through observation, imitation and emulation – as those learned behaviours that do make a critical impact on survival, like what foods to eat or when to run away. And the more complicated the behaviour needed to thrive or merely stay alive, the more we desperately need this kind of social

---

* And, in a shot that should have won the Natural History Museum's Wildlife Photography prize but inexplicably didn't, looking bored while playing with an iPhone they've nicked out of some tourist's hand.

transmission of information, because you just aren't going to learn how to be a successful macaque by falling in a pool. You need to know what food to eat, where to find it, who to groom and who to avoid, and how to fit in. You need to learn macaque culture and, yes, that might mean learning how to hot tub.

Social learning is the absolute key to primate success and in no primate is this more true than ourselves. We have such a terrible lot of culture and we use it for nearly everything that any sensible animal would revert to hardwiring for: finding food, finding mates, surviving. The entire latter part of this book looks at how we try to stuff culture into our offspring to give them the best possible chance at life, because this is what we do with our long childhoods. But before we get there, we should understand how this stuffing happens – the processes of social learning. We need to look at the way social learning occupies the various stages of our growth, because these are the evolutionary levers that we have been hanging all of that cultural transmission from and we have definitely been manipulating them.

The hallmark of infancy in primates is dependency on breast milk for most or all of the animal's dietary needs, and in primates it is marked by close contact with the primary carer. All of life is filtered through this one being – usually the mother, but – as we saw in Chapter 12 – occasionally the father or some other alloparent. Infancy might be seen as the period where the baby primate is glued to their one primary teacher, observing and learning. Infant primates in fact spend a vast amount of time doing what is called 'peering', which any human parent may well recognise – bringing their faces right up to intently focus on a parental activity. Chimp babies have been seen to do more 'peering' at mothers eating new foods or eating more complex foods, and generally do more 'peering' in advance of making their own attempts at the trickier things they've seen their mothers do. Much of the social learning passed on in primate early years seems to concern the 'what' and 'how to' of eating, actually, a critical aspect in the business of survival of course, and a skill the animal will need when they are less directly provisioned. The tight dyad of mother and baby (or, with our proud baby-wearing marmoset, father and baby) forms the ideal environment for this focused acquisition of the skills that are going to keep it alive once the milk stops.

A variety of species have demonstrated exactly this kind of dietary training under observation, from vervet monkeys' mothers being trained to eat corn of a specific colour and passing this information down wholesale to their offspring, to more complicated processes like how to smash nuts open. An extreme example is that provided by the female chimpanzees of Gombe National Park in Tanzania and their clever daughters. Gombe chimpanzees are keen fisherfolk, fashioning tools out of blades of grass to 'fish' out delicious, protein-rich termites from holes in their mounds. Mothers fish the termite mounds either by themselves or in small groups of closely related females, and it is from these groups that infant chimps learn the skill. Daughters are much quicker to pick up this task than sons even though both sexes spend equal time hanging around the termite mound, which prompts questions about exactly *how* this differential acquisition of skills happens. There isn't any suggestion for any basic hard-wired differences in chimpanzee learning between males and females,[*] but

---

[*] Having not gone through the 'gendered brain' convolutions of mid-twentieth century neuroscience to the same extent that humans did.

observers noted that daughters spent more time actively observing and imitating their mothers, while sons spent ever so slightly more time larking about. And when they grow up we find daughters have fishing techniques that much more closely resemble their mothers' than sons do.

Social learning then, is not a one-size-fits-all experience, even in our primate relatives. Tool use seems to be a situation in which females serve as the main teachers and curators of skill. Hunting is a behaviour once regarded by anthropologists as a precursor for our own species 'man the hunter' take on protein acquisition, and often assumed to be absolutely and unavoidably the realm of the male animal. Observations of savannah-living chimpanzees from Senegal, however, have shown that females (and youths) utilise stick tools to hunt, while adult males rely on speed and strength to bring down prey. It seems that chimpanzees, like humans, do grow up to have some gendered practices – but, just like in humans, these are only gendered in specific group cultures; not everyone teaches their daughter how to skewer a baby galago on a stick.

This cycle of observation and imitation has been argued to be the hallmark of primate social learning and it is not limited to infancy – this behaviour follows the growing primate right through its life. What changes, once the animal in question is weaned, is the number of potential teachers. As primates with the teeth to eat nearly anything in a world full of inedible things, we don't so much learn to forage for ourselves as we learn what our mothers, friends and other adults eat by observation and imitation. The most famous example of this comes from the rather fabulous chance observation of the seaside-dwelling troop of macaques on Japan's tiny Koshima island. In the 1950s, researchers were starting to feed the monkeys, dropping off bits and pieces for them to eat and watching their behaviour. One day, a local elementary school teacher happened to see something utterly unexpected: a monkey called Imo took a sweet potato, walked down to a small stream, and washed it in the water. Within a handful of years, Imo's siblings and mother were all washing their potatoes, and not long after the potato washing evolved into a whole troop of monkeys who dipped their potato snacks in the sea, essentially inventing sweet-potato fries (though admittedly without the frying).* Imo's sweet-potato lesson had spread through her peer group, a clear example of social learning, but one that went sideways rather than coming down from a parent above.

Early childhood marks the first big foray into a wider world of sociability in most primate species. Given the freedom from being physically linked to their primary carer, younger children start to expand their social circles. In gregarious primates, where there are hordes of relatives or allies to meet and learn from, this is when the child begins to spend appreciable time socialising with other animals, watching and learning. Imitation still plays a large part in this approach to learning, but now the primate child is free to select new models to copy. There is some evidence in non-human primates that particularly skilled examples are actively sought out – capuchin monkey children, it seems, spend rather more time staring intently at the best nut-crackers among their troupes than the merely so-so. Primates pick who they want to learn from in this way, choosing older individuals rather than younger and, when

---

* This isn't even the only time sweet-potato washing has been invented in primates, or even in Japanese macaques. There was another troop that used to get hot sweet potatoes and learned to drop them in water to cool them down.

it comes to behaviours that are largely male or largely female, selecting a specific sex to observe and imitate, like in the stabby chimpanzees. The important thing is that the pool of potential teachers – and therefore the pool of potential skills – is expanded. It doesn't matter if your mother doesn't know how to salt her sweet potatoes if your best friend does.

So, what about this critical bit of growing up needs so much time that we have to extend our childhoods just to take it all in? Like other primates, we start to pick up the skills needed to make friends with others as we detach from our mothers in the last stages of infancy. Our juvenile period sees us exploit these networks of friendly faces for skills and information, but quite a lot of the social learning we do is no different from our primate relatives: we sit and we stare. Or we stand and we stare. Ethnography of people who live in small groups and have to learn skills such as game hunting and foraging has revealed an impressive absence of what we victims of formal education might recognise as 'teaching' when it comes to how to make a stone tool or how to weave a basket. Even the process of making complex stone tools, ones that require multiple steps to complete, might theoretically be learned entirely through passive observation. It is incredibly difficult to shape a pointy rock in such a way that will hit a moving target on the end of a stick, even more so to do it so that you are likely to not only survive the encounter but get a meal out of it as well, but you can learn how to do it by watching.[*]

---

[*] There are numerous archaeologists who have pretty fun sidelines doing things like recreating stone tools to learn about their manufacturing techniques and it is notable that the vast majority of them specialise in big easy hand axes.

However, there are things our kids need to learn that they don't get just from watching. Primates need something extra to survive: society. That means that we need to learn to get along in whatever society we have. In many primate societies, adolescence, the tail end of childhood – the one we shouldn't really have, the bit where we might technically be able to reproduce but still insist on playing around – triggers a rapid change in social situation and sees the dispersal of young primates out into the world beyond their birth groups. For primates like the fantastically wide-bearded langurs of Southeast Asia, the arrival of new males into their birth groups comes with severe consequences – older male juveniles and young adolescents will have to move on, or risk injury or death. For groups where males or females are expected to up sticks and relocate, either hanging around on their own or in small packs of males, this means that just surviving childhood requires an adaptation to new social circumstances.

Moving group might mean the loss of status and signal the need to reorient social behaviours accordingly; it certainly means that the local knowledge a primate gained from its birth group needs to be adjusted to a new set of trees to sleep in, rivers to drink from and places to acquire food. For many primates, moving to a new group means learning the local habits: female chimpanzees have different grooming traditions, for instance, and as you can imagine a newcomer will never flourish with an out-of-date grooming routine. However, this adolescent dispersal – alongside the movements of individuals that happen less frequently in adulthood – marks a unique opportunity for the transmission of primate knowledge beyond the borders of a single resident group. Young primates who learned termite fishing or rock-playing or hot-tubbing can now carry these innovations to their new groups. Adolescence is the period we spend consolidating social capital into the relationships we will need to survive.

Chimpanzee males, who will need to depend on each other for help when it comes to activities as varied as mating, support against other males and hunting, have spent their little-kid years chomping at the bit to hang around in groups of older males and are suddenly free to indulge. Likewise, most great-ape females choose female relations to socialise with right up until they begin mating, and if they are going to remain living nearby they can begin to form alliances of support and protection. Even for animals that will stay in their own birth groups, the transition from juvenile to adolescent is when the kinks get worked out of eventual social status. Female baboons, for instance, will most likely spend their lives in the same groups they were born in and will generally be treated with the same respect as their mothers. Starting even before they hit the beginnings of sexual maturity, however, they will begin to challenge that assigned hierarchy, doing their best to menace other lower-ranking females and therefore move themselves up the ranks. All of us primates invest in our children by preparing them to live in society, humans no less than baboons. If you want to give your children – and their children – social capital, you've got to do the work of learning and maintaining social relationships. We aren't the only species that knows who its friends are, of course. But we are perhaps the only species that teaches our children who they are.

Children among the Qiqiqtamiut of the east Hudson Bay area in the 1960s, for instance, were 'taught' from a very early age to recognise kinship terms by means of a sort of peekaboo game that would be played out whenever the child was in amenable company. Someone would name a kinship relation – aunt, say, or uncle – and if the child managed to make eye contact with the correct individual, there would be a round of congratulation and positive reinforcement. In the small settlement of about 35 people where this was observed, by the age of about a year and a half children could identify every member of their group and their relationship to them. This is the sort of skill humans are willing to waste words on. While observation may be a useful teacher for some lessons,[*] ethnographic evidence points to the importance of other modes of social learning like storytelling, proverbs and myths that offer children insight into the social world they will interact with. Words, it seems, are how we teach our children the vast amount of things they need to know – and this is probably as true for our species 100,000 years ago as it is now. Language is how knowledge of the natural world, with its infinite complexities of animal movement, seasons and landscapes, gets passed on. It is the tool we use to capitalise on our social capital.

What does a child need to know to be human? There is an ever-present danger of trying to define 'childhood' and ending up in a palaeo-place; trying to scope out what we need to learn in childhood for risks creating a monolithic and unchanging mythical ideal that Jung would be proud of. But what exactly would we attribute to this Jungian archetypal Ur-childhood? People say a lot of things about childhood, how do we know what is nonsense and which

---

[*] Hitoshi Anezaki, one of the last Ainu bear hunters, recalls just such an experience. 'I really think that the bear is my teacher. When I was young I didn't know anything about hunting. Then I decided to follow after the bear in the mountain, walking through the mountain as the bear did, taking a rest as the bear did, tried to feel and think as the bear did. I imitated everything that the bear did. Through those experiences, I obtained all the knowledge on nature and hunting.' In *Anezaki and Katayama* (2002), translated by Terashima *et al.* (2013).

are evidence based?* Well, locked into the discards and debris of the archaeological record, is where we can finally access the real palaeo-childhood of our ancestors, all those thousands of years ago. And there wasn't just one childhood, obviously, there were hundreds of thousands of them: a huge number of ways to make a living have presented themselves in the few hundred thousand years since we became us. Childhoods were not all in caves, or set in the backdrop of an African safari brochure. For at least the first half of our 300,000 years, all of our children were probably in Africa, riding out the increasingly cold Quaternary period with its glaciers and fluctuating climate. Some places were probably more hospitable than others, leading to people pooling together in unprecedented ways. The accumulation of so many humans together in these particularly not-terrible areas – technically, they are known as refugia – may have done the same thing cramming too many primates in one place always does – force adaptation.

While things above the Saharan belt were cooling down, life on our home continent became humid … and inventive. A whole new raft of activities for staying alive appear in what is generally called the Middle Stone Age, around 280,000 years ago, and the exact same thing that happened to primate sweet-potato lovers and hot-tubbers happened to us – it just happened on a grand scale. Our repertoire of sharp rocks starts to include points that can be hafted and thrown at more difficult prey; 'to catfish' is invented as a verb for the first time when we come up with harpoons. Needles and awls to string together the clothing needed to deal with an inclement age appear in the archaeological record. In the pleasant places like the South African coast,

---

* Step one, buy this book.

whole new decorative traditions arise around 80,000 years
ago, with beads, engravings and ochre paint industries to
spruce them up. We made all of these clever discoveries,
and we shared them with our friends and families, and
because we get around our clever ideas got around too.

These new life skills form the base rock that human
innovation and expansion depend upon, but are they really so
different from the 'discovery' that sweet potatoes and indeed
everything could, and possibly should, be salted? Well, yes. No
one wakes up one day and decides that they're going to start a
continent-wide trend for pounding particular reddish or
yellowish rocks into powder, which is how you get ochre, the
number one pigment of the last few hundred thousand years.
It's really a lot of effort, for one thing – first you must learn
which rocks are suitable for grinding up, how to grind them
and – this is even more important – you need a reason to do it
in the first place.* If pigment – paint – becomes an important
part of your social life, you really do have to ask why.

And here is where we start to see where our human
capacity for communication takes on its outsized importance
in the way our children learn. Paint, decoration – these are
symbolic forms of communication that convey important
meanings, but those meanings are things that society
teaches you. Just like the Qiqiqtamiut name-game, someone
has to explain the symbols we cover ourselves in, what they
mean, who they mark us out as. The same is true for beads,
and probably a host of other things that don't survive the
millennia like hairstyles or clothing. The way we decorate
ourselves can be an important symbol of who we are,

---

* You don't actually have to be an anatomically modern human,
however. Neanderthals seem to have exploited red ochre from
around 250,000 years ago. For those considering a degree in the
visual arts, please note it wasn't enough to save them.

where we are from and what we're about; whether it's
ochre-stained shell beads from Blombos Cave 80,000 years
ago, or football fans in face paint in the 1980s. So not only
are there an immense number of practical skills to learn in
a human childhood, there are new sets of social meanings
that any kid is going to have to parse if they are going to
get along in life.

Children are a storehouse for the technologies and skills
that allow early humans to do so many wonderful things.
While the investment into bodies as hard physical capital
remains unchanged from our primate past – *Homo erectus*
parents fed their children too and taught them the skills to
shape a stone – we can wonder about the amount of
investment in social capital setting our species' apart. One
of the reasons for this is that there are just so very many of
us, and we have been long-distance wanderers for a long,
long time. Thinking of the relative-naming game, you can
imagine that it would take a *lot* longer for a kid to learn the
names and blood relationships of 300 people versus 30. If
you are looking at humanity's meandering, climbing
population, you are also looking at the climbing numbers
of complicated ways those humans are connected to each
other. Many researchers have argued that the existence of
so many people requires convenient new shorthands for
relating people; a sort of magical reimagining of kinship
that allows shared ancestry to group people into different
sets, or lineages, so that they can still be properly 'named'
in whatever your society's equivalent of the name-game is.
And when you cannot articulate the relationships between
people who have to live in close proximity to each other,
you are going to need symbols and narratives to do the
introductions for you.

The stories and symbols that adults use to teach children
are all serving to build up a child's social capital – their

networks of friends and family that recognise them as kin and will invest in them, well beyond the capacity of any individual monkey. The stories and symbols might vary hugely, but the critical point remains the same: human kids need teachers. They need the social networks of their parents, and their parents have got to put the effort in if they want their kids to do well. But, of course, standing around staring at mama making hunting spears, or learning how to respond to the question 'is that my paternal uncle', aren't the only things we fill a childhood with. We have all of this space between infancy and reproduction to fill — and what's left to fill it? Well, the answer, of course, is child's play.

## CHAPTER EIGHTEEN

# Girls and Boys Come Out to Play: Learning the Easy Way

Girls and boys, come out to play,
The moon doth shine as bright as day;
Leave your supper and leave your sleep,
And come with your playfellows into the street.

If someone were to ask you, wildly unprompted, what it is that children *do*, you already know how you would answer. Children, as we all know, play. Childhood is the time an entirely new mode of social learning comes into being – the potential of learning through play. Play is, essentially, practice for the rest of life. Play is anything that involves you knocking around without a clear goal or purpose, even though it is burning up valuable energy or potentially exposing you to danger. And, despite the obvious disadvantages of wasting time, food and luck, we all do it. Animals from across the spectrum indulge in horsing around. We don't actually entirely know why; play might have arisen in so many species in order to boost developing motor skills, bond with peers or simply as the only possible response to existential boredom. Or it might have a more critical function entirely: learning.

Play definitely does function as a method of learning. It has adaptive benefits in terms of introducing unexpected scenarios – being knocked over by a frolicking fellow foal, for instance – that require some mental or physical flexibility to address. Animals from marmots to horses that indulge in more play are frequently observed to be better at acquiring motor and social skills, while those that either play less (or are deprived of the opportunity by behavioural scientists) end up taking longer to achieve physical competencies and can be rendered almost completely socially incompetent.

We are no different. The scrabbling and larking around that characterises many primate childhoods leads to improved motor skills; young chimpanzees that play more are quicker to master important skills, like riding their carer like a horse. These skills can come at the expense of general growth because trying (and failing) to climb up on your mother is a calorie-and-mother-exhausting business. Nonetheless, play still seems to be an adaptive behaviour that has radiated out into the childhoods of nearly all of our relatives. Primates in particular need to play, because they are not just learning how to sit side-saddle; there is so much more to primate growing up than learning to move around. Primates need play to set themselves up socially; to figure out their place in their group and how to keep it (or improve it). We need play because we need to work out how to monkey: research has shown, for instance, in young macaques that isolation makes them socially awkward and overly aggressive as adults. They never learned to play and so they never learned how to adult either.

The nature of play also tells us something about the societies we are trying to adapt ourselves into. For instance, species that put a great deal of emphasis on social relations are early adopters and encouragers of social play, sending infants out to bond with other group members. Chimpanzees accomplish this kind of delicate diplomacy through sending kids off to play games with other troop members; letting aunties and uncles and friends introduce games such as the 'airplane' – propping up a spread eagle baby on the hands and feet of an adult lying on their back – a game that may be very familiar to human readers. Play starts with primary carers for infants, but as baby grows, it rapidly expands to include other adults, depending on the rules of the primate society (and the tolerance of individual primates). Playful ape children grow up to be playful, tolerant ape adults, something that assists them in making adult social bonds, defusing social tension and even getting a sexual partner. This is exemplified in the extreme by the deeply unserious bonobos, who rely incredibly heavily on play interactions to manage their societies.

Play with other young primates offers a uniquely horizontal avenue for social learning. The propensity of children to come up with novel entertainments for themselves seems to be widespread throughout the primate clade. As we saw in the previous chapter, innovation in behaviour can lead to the chance discovery of some pleasant new skill – such as hot-tubbing or sweet-potato washing – that ends up being imitated and then transmitted both up and down generations. But as one of the most prolific wild-living species of monkey, macaques have also been observed in all sorts of strange new inventive behaviours just for the fun of it. A young female macaque discovered in 1979 that stones were fun to play with, prompting 30 years of study of exactly how this behaviour spread throughout the macaque social milieu. The answer? Friends. Much like the juvenile who learned about hot tubs, the game of stones was first spread through playmates, until eventually those juveniles grew up and had babies of their own – who they taught to play with stones.

There are age-related changes to play, with infants and very young juveniles obviously unable to keep up with more complex games. Unsurprisingly, it is the long period of childhood – after infancy, but before the business with reproduction starts – when play is the most seriously pursued, and play expands to fill more and more time as it becomes a social enterprise. While mothers play with their infants and younger children engage in all sorts of play on their own – from amusing themselves gambolling about, to picking up and putting down new and interesting stones, through to cautiously interacting with their peers – if they are allowed, it is in the later stages of childhood that play really comes into its own as a social tool.

So, how do human children play? We are all aware of the spectre of video games and screen entertainment, lurking in the imagination of Tipper Gore and other cultural refuseniks as the gateway to the destruction of all human society.* We may in fact already be doomed, with many children of the 1980s and 1990s now dangerously desensitised to human life by exposure to two Italian plumbers and some suspect mushrooms. This sounds facetious, but play is still a grave concern. We recognise as a species how important play is, and you can tell because we agonise over it. In the parenting cultures of recent times in affluent places, there is disdain for anything but the 'best' play; mothers (always, the mothers) must be fully engaged in creative activities that engage motor learning, category discrimination, and other skills that for some reason can only be acquired by sacrificing your day job and your carpets on the altar of 'appropriate play'.† Play is teaching our children how to get along in their world – screw it up

---

* Note to the youth: Tipper Gore was the wife of US Vice President Al Gore in the 1990s and boy was she worried about bad words and good games.
† My kid's day care has an app that gives me a blow-by-blow description of the key skills each play activity she participates in imparts. I don't know how to tell them I just want to see the finger painting at the end, and could care less how familiar she is with 'colour and texture play'.

and you've doomed them for life. This is kind of heavy going for something that is meant to be fun.

So how we are 'supposed' to play? What have we evolved to do, and why? Well, as always, there is a quite shouty school of thought that would like play to be as stripped back to basics as possible. While I am not currently aware of anyone marketing 'palaeo play', it can only be a matter of time before someone tries to sell you an Ice Age activity. The thing is, play is everywhere. It is ubiquitous. Play happens alone or among peers, with objects or without, but it happens in every culture. The only thing that changes is *what* play is about. What we played with in the past may not be what our children need to play with in order to learn *their* society. Play doesn't even remain static within a generation, let alone over the deep time we are talking about for human history. I personally remember being fascinated by a board game in the late 1980s that featured a fake landline phone that would play tinny recorded snatches of conversation and I haven't even had a landline in the past 20 years.[*]

Generally, children seem to spend more time playing at tasks that they are likely to end up being asked to do as adults. For example, in several societies where hunting holds an important role in subsistence, children are given modified 'child-sized' versions of hunting weapons and tools to play with from an early age. So, not only might children get different toys based on age and ability, but play also changes depending on what those children are expected to grow up to be. In societies where the means of survival depends on pretty biologically deterministic roles, this means differences in play between biological sexes as well.

---

[*] Though perhaps it should be brought back. I cannot imagine any activity more guaranteed to set someone born after the millennium on edge than speaking on the phone.

Play, after all, is training, and it isn't until very recently we have had a society where all roles are (ostensibly, if not in practice) open to all people. Even now, despite the relatively wide array of means of subsistence and opportunities for work now available to humans, this gender split in play is *still* found near universally. Boys and girls in a wide array of societies prefer the company of their own gender for play after the age of about two and those preferences solidly entrench after they are about six. Child's play mirrors the reality of our societies; whatever social rules we speak out loud, play tends to unearth the unspoken stuff kids rightly realise they're going to have to learn.

Interestingly, among the so called 'egalitarian' small communities that make their living through foraging, however, there seems to be less of an emphasis on gender in determining who plays with whom, at least among the youngest children. This suggests that the absolute determinism of biological roles that we imagine for the past – think 'man the hunter' nonsense* – has less impact in relatively small groups making their living doing a little bit of this and a little bit of that. Comparison of two Bofi cultural groups in Central Africa – one habitual farmers, the other habitual foragers – showed that while children definitely played in groups that would be related to their eventual social roles, there was an earlier and more distinct separation by gender in the farmers, whose adult lives are more circumscribed by gender, than in the foragers. Not only that, but farm children get less time for play overall. Watching the social lives of Hadza and also Aka forager children, it seems clear that living in smaller groups, or

---

* Now wipe that entire trope from your brain. Perhaps replace it with the image of heavily armed, murderous she-chimps from the previous chapter.

relying on foraging and hunting, equates to more time spent in play than in the more constrained economies of farming communities.

Reconstructing the play of the past is actually incredibly difficult, because we don't have access to much of the hard physical evidence of human behaviour. Which is a pity, because play is precisely the sort of thing that would tell us, far clearer than modes of production or technology used, what people in the past were training their children to be. Unfortunately, anything made out of organic materials like reeds, wood or fur – which is almost everything in the past – will have disintegrated (mostly) out of archaeological reach by now.* There are, however, traces and suggestions of even the smallest lives in some of the material culture that has survived down the years. These artefacts allow us to examine some of the expectations of childhood as the period where the first steps are made towards the acquisition of skills needed to survive as an adult and, more importantly, let us guess what it was we have been doing with our childhoods down the years.

It is probably clear to most readers that in our Upper Palaeolithic past, skills like hunting and tool-making would have been prime targets for play. But what do we actually see of children's play in the past? Anthropologist Michelle Langley has pointed out several of the difficulties in picking out the most obvious correlate of play – toys – from the archaeological record. Toys may have been made of materials that did not survive the long trek down the years, rather than the stone or bone materials that adults would have needed to make proper functioning tools. Children might

---

* Mad props to the archaeobotanists and phytolith specialists who study the microscopic fragments of organic material. That's even fiddlier than staring at teeth.

have played with discarded adult tools, blunted or cracked points that wouldn't be distinguishable from other refuse once they'd gone into the ground. Decorative items might have cracked or eroded and then been handed off to children, or they might have been given to children and broken through play – or they could have just been discarded, and in all cases the artefact uncovered would appear identical to the archaeologist. And, of course, there is no way of knowing the intended user of any artefact, even if it is 'child-sized'. One of the most vexed questions in uncovering play in the past is we aren't actually all that sure where child's play stops and the imagined world of adults begins. Humans are entirely capable of making almost anything in their material world stand in for something else; this is symbolic thinking and it is a type of thinking we excel at.*

Whether they are used as the basis for imaginative play or didactic learning, there is no reason to think that children were excluded from the important symbols and stories of their cultures. Storytelling may have been one of the most important ways of learning about the world in the past and, in the last 40,000 years or so, it started to emerge into the physical world. Representations of *things* start to appear on cave walls and the sides of rock shelter, with the oldest known cave art a sort of half-hearted pig currently dated at between 40,000–50,000 years old and scribbled on the interior of a cave in Indonesia. Australian rock art may date back at least that far as well. In Europe, the famous caves of Lascaux, Altamira and Chauvet are later, but have been better known for longer in popular culture.† These depictions of animals

---

* Possibly occasionally to our detriment. Bourdieu makes for some heavy reading – and sometimes a cigar is only a cigar.
† Despite the fact that Australian Aboriginal art is part of a tradition that continues to this day and has actual artists around one could talk to.

and humans, particularly when they can be attributed to 'scenes' of action like hunting, have been held up as possible vectors for the transmission of important cultural information. And no one needs information like a child needs information.

We are fairly sure that these artistic endeavours must have at least occasionally had some active role in community life. In the depths of Lascaux cave, there is a line of animals headed by a charging bison. Littered along the floor are discarded stone lamps that would have burned animal fat and provided a flickering light that, when played across the rampaging animal figures, could have given the impression of movement. In Chauvet, there is a series of depictions of a rhinoceros head in different positions that, if they could be animated, would show the animal lowering its horn. And perhaps most charming of all is the discovery of little bone discs with perforated holes, originally interpreted as pendants or buttons. These discs have two sides, usually with the same animal depicted in different poses. These may have been 'thaumatropes', images marked on discs that could be rapidly spun when mounted on a bit of string, transforming a standing deer to an arrowshot doe with a pull of the string. While these moving displays may have been intended for an all-ages audience, we know that children would have visited this symbolic realm as well. In many of the caves and shelters that preserve rock art, we also see the marks of human hands – perhaps signatures, perhaps symbols – outlined in pigment and left alongside the animals and the action, or as fingers trailed through soft clay into shapes and patterns. And among the many adult hands, we see the indelible imprints of children.

Of course, there is an eternally popular type of symbolic representation – one that is still incredibly common in the lives of children today – the humble doll. Dolls, you would think, are simple. A doll is there to be played with as though

Figure 18.1. *The Laugerie-Basse 'thaumotrope' is a bone disc that could be mounted on a string to animate the two images on either side. After Langley (2017).*

it were a real human – and usually a real baby human at that. Children the world over mimic the baby-caring behaviour of the adults around them, whether they live in small mobile groups as we think children in the Upper Palaeolithic mostly did, or in the suburban cul-de-sacs of the developed world. We have plenty of ethnographic evidence that modern children play with dolls made of just about anything – even amid the commercial excesses of the late twentieth century's plastic-toy epidemic, I was taught that a few twists and a knot would make a perfectly serviceable doll from corn husks.* Aka kids are perfectly capable of similar feats with banana shoots. And dolls – or at least, human figurines – are one of the very first types of artefacts that aren't sharp and pointed for killing things that we even know of.

We find 'dolls' starting as far back as 40,000 years ago, but – and this is interesting – we are rather less inclined to think the few precious decorative objects that have come

---

* There is also an American folk tradition of making doll heads from dried apple cores; the resulting dolls go a long way towards explaining the presumably related American folk tradition of being absolutely terrified of dolls.

down to us from that long ago are for anything as banal as child's play. Even though we've got grubby kiddie handprints in the fancy art caves, we are determined that the little figures of the deep past are for adults only. The tiny, less-than-6cm figure carved into a bit of mammoth ivory and dropped in the German cave of Hohle Fels is that of a woman with hugely exaggerated bust, belly and butt; she is featureless, headless and footless, which turns out to be a typical look for female, or probably female, figurines found all over Europe and Siberia during the Palaeolithic era.[*] Also from Germany around the same time is an impressive standing figure carved in mammoth ivory identified by its discoverers as half-man, half-animal, and known as the Lion Man.[†] The painstaking work of carving them and the potential symbolic significance are thought to rather outweigh the potential amusement value for children.[‡]

---

[*] I am not going to call them 'venus' figurines. The first ever 'venus' found, the *Vénus Impudique*, is a girlish figure whose discoverer was attempting a comparison with the statuary of classical times. However, the common adoption of 'venus' figurine for the fat, hyper-sexualised statuettes comes from comparison to the body of Sarah Baartman, a Khoikhoi woman who was displayed to the public as an archetype of 'primitive sexuality' in what must be one of the least edifying, most racist, misogynist and generally 'othering' moments of colonial anthropological 'science' in a field that, to be fair, is not lacking for them. So, no. Campaigner Victoria Herridge has suggested introduction of the term 'doni' instead, which has the advantage of being taken from Jean Auel books and therefore reassuringly fictional.

[†] Despite the reams of paper expended on discussing the symbolism of this very first chimera, the actual figurine looks like nothing so much as a perfectly normal cave bear standing on its hind legs.

[‡] One suspects these people have never actually had to entertain children, or they would realise any amount of effort is acceptable.

Likewise, other miniatures are interpreted as models of social organisation created for ritual purposes, rather than little models of the world around them created for play. Little clay figurines with detachable heads and moveable parts found in the villages of the earliest farming communities some 12,000 years ago are assumed to have some great ritual significance, no matter how fun they look. And while the museums of the world are full of Greek and Roman figurines that look like they'd rival the pricier end of the modern plastic doll market, we actually know they are not for fun. A bit more than 2,000 years ago we see fabulous jointed clay dolls with little moveable arms and legs appear in Greece; from traces left behind in the soil we can guess that these were probably dressed and painted as well. However, it is where we see them gives the game away. These 'dolls' are found in temples and in graves alongside adults; this tells us they performed a more grown-up symbolic role as effigies and votives. Most of the things that look like dolls across the ancient world were probably intended as votives, offerings or were parts of household shrines serving a symbolic purpose rather than embodying the interior imaginative lives of children.*

However, children of the ancient world should hardly be thought of as being bereft of playthings.† The culture of the Indus Valley some 4,000 years ago, with its seat at Harappa, produced considerable evidence for a better balance between the artefacts of serious business and little

---

* At least when the grown-ups were looking.
† As your grandparents have no doubt told you, it is possible to be amused for hours with a piece of string, given sufficient time has passed between the moment of recounting the amusement and the actual experience of being a bored child.

objects that we could associate with play. At the site of Mohenjo-daro in modern-day Pakistan, we see little clay birdcages, too small for any real bird; we also see clay whistles shaped as birds. Miniature carts, wheels and pots, made with varying degrees of skill, are also argued to be testament to children's play. There are marbles, and clay balls and figures, and polished stones that could be game pieces or slingshot pellets, or any number of other amusements. There are even dice, which seem to be a very long-lasting part of human play. There are early examples of dice from what is now Iran more than 5,000 years ago – and Skara Brae, on the tip of the United Kingdom's remote northern island of Orkney, only a few hundred years later. Play is not so different, no matter where you are. Just like in southern Asia, we see that children's graves from burials at the 3,000-year-old Great Pyramid at the site of Cholula in modern-day Mexico also have whistles and flutes, alongside ceramic balls and figurines.

Though they may be just as fun for adults as for children, even board games are of considerable antiquity – at least four 3,000-year-old sets of the ancient Egyptian game Senet were found in celebrity teenage pharaoh Tutankhamun's tomb. Where I excavate in the northern fringes of Mesopotamia, at Başur Höyük, we see what are perhaps the earliest game pieces yet recovered – almost 50 small pieces,* lovingly carved in precious stones more than 5,000 years ago. There is the Royal Game of Ur, or

---

* One of the highlights of having the world's earliest abstract game pieces on your site – because, let's face it, people have been playing with sheep knuckles forever – is the joy of arguing long into the night about whether something the size of your thumb that's been underground for millennia is a pig or a hedgehog (I'm Team Hedgehog).

'Twenty Squares' first found in the Royal Cemetery of Ur
a few hundred years after the game pieces were buried at
our cemetery in Başur; the characteristic board shape
shows up across the ancient world for millennia, however,
from Crete to Sri Lanka, and people actively remembered
playing it in the 20th century on a kibbutz in Israel. My
personal favourite example of the ubiquity of the game is
the board carved into the otherwise terribly imposing
gate-guardian sculptures that patrolled the entrance to a
palace at Khorsabad, made almost 2,000 years after the
version found at Ur was put into the ground. Under the
watchful eye of the *lamasu*, the winged bulls that served
the ancient Middle East as spiritual guardians, the more
human guards stationed below the majestic, 2m-high
figures scratched themselves a game board right into the
stone. You can still see their handiwork, a millennia-old
testament to boredom at work, today – the *lamasu* figurines
were 'collected' by the British Museum and if you check
the small bench under the right-hand one's belly, you'll
see it yourself.

   Children also act out their caring play on living
creatures, whether this is beneficial to those creatures or
not. Pets are a frequent feature of many forager
communities, particularly juvenile primates from species
where the adults are more commonly recognised as food.
Hadza children, for instance, use infant galagos (here
more appropriately called by their common name, 'bush
babies') as dolls, while Aka kids also play with infant
monkeys. Perhaps the most prolific pet-lovers in all of
humanity come from groups who make their living
largely from foraging in the mix of rapid development,
agricultural encroachment and forest landscapes of the
Amazon. The Guajá of Brazil, for instance, are reported
as keeping about 90 baby monkeys as pets in a group of

about a hundred people, which is a very high rate of baby monkeys per capita. However, the Guajá know what every pleading child with an eye on a puppy for Christmas does not – pet keeping is mostly for adults.

We do know that at least one kind of pet featured in children's lives in the past 30,000–40,000 years of human history, and it should surprise no one that it is the dog. Dogs have been by our side for tens of thousands of years, even as our childhoods expanded from the world of finger-painting and magical symbols to the plastic-choked consumerism we see today. In Chauvet cave, known for its fabulous depictions of Ice Age beasts, there is a long-since-hardened path trodden through soft clay by the tiny feet of a child, who would have been about the size of a modern six-year-old, made more than 20,000 years ago. And alongside their unhurried path? Set forever in clay, the heavy tread of a large canine companion.

Most importantly, all of this business of learning to be human takes a *lot* of time, whether it is learning the rules of society or trying them out through play; it doesn't even necessarily end when puberty hits, or adolescence turns into full-throated adulthood. But childhood is when we turn the tap on. Social learning is the key to surviving in our cultures, and parents invest heavily in giving their children the time and opportunity to build up their social capital just as much as they invest in building up the actual embodied capital of skills and healthy growth.

Overall, the business of being a child has not changed, only the operating hours and the working conditions. But our success breeds new problems. More and more children are born, and the world begins to change again as society stretches itself into new contortions to try to keep them all alive. After hundreds of thousands of years of kids just being kids, those contortions begin to place some children *here* in the order of the world, and some children *there*, and all the while their parents struggle to make the investments that will keep their children *here* – whether that is on top of society or just on top of the soil. In the next chapter, we will see what happens to childhood when the world starts to demand a lot more capital for a child to survive – not just the physical investment that all animals make, or even the social capital that our primate clade is famous for, but real cold hard material capital. Because in this world, you've got to work for a living.

## CHAPTER NINETEEN

# Jack and Jill Went Up the Hill: the Hard Work of Childhood

Jack and Jill
Went up the hill,
To fetch a pail of water;
Jack fell down
And broke his crown,
And Jill came tumbling after.

E veryone knows that an older child is a more useful child. They've got the motor skills to do things, but lack the social (or physical) stature to refuse to do them.* We saw in the previous chapter that the nature of play was subtly different among free-range children and children living locked to the land. But what about play that achieves things – or, as we might think of it, work? We have seen how societies pour resources into their children in the previous chapters. Well, now that they are standing on their own two feet, feeding themselves (just about) and full of complicated symbolic ideas, that kid they built is ready to give a little something back. In the game of being human, no one gets to sit on the sidelines. Being a kid is all about learning how to adult, so our kids have to work, because that is what grown-ups have to do. Or is it?

One of the most tantalising ideas in the history of the study of humans comes from the suggestion by

---

* Interestingly, this is not always true. There are cultures that find it absolutely unconscionable to demand a child go run an errand or some other thing they don't want to do.

anthropologist Marshall Sahlins that, actually, being a
human doesn't have to be very much work. He published
his compilation of ethnographic insight into the lives of
modern-day (or at least, his day; we are talking about the
middle of the twentieth century) foraging people and
suggested that, actually, a lifetime of foraging left quite a
lot of free time for loafing around. Reading studies by
ethnographer Richard Lee that suggested Khoikhoi spent
no more than 15 hours a week 'working', he postulated a
very different sort of employment than the one we see in
heavily capitalised societies today.

The idea that even foragers making a go of things on the
absolute edges of habitable human space, pushed to their
limits by encroaching farming, herding and industrial
modes of production, could be having a very chill time is a
lovely one. It just happens to also be very unlikely and the
problem is this: what a (male) Canadian ethnographer
would recognise as work in 1968 is not necessarily what
everyone else would recognise as work. Lee calculated
work *only* in men. Because, presumably, only men work,
especially if you define the only real work as 'running
around hunting animals'. The constant occupation of women
with the billion things that need to be done to make a
society function then is not really *work*. Except of course it
is. Staying alive requires effort. When it comes to things
we think of as jobs, there have always been 'special' tasks,
even when everyone was pretty much doing the same thing
to survive. What people did for a living in the thousands
and thousands of years of the Palaeolithic may have covered
a wide range of tasks (tool-making, child-wrangling,
trapping, shaman-ing, gathering, *etc.*), where some people
did more of one task than others. But everyone in a society
that is able to contribute is 'working', even if it's just picking
up some tubers for dinner or telling a bedtime story.

Who needs to work at what is something that does have to adapt as we change the way we live, however. Ethnographically, we see children from non-mechanised farming communities are far less capable of supporting their own nutritional needs than the children of foraging communities. In snack-rich environments, a competent gathering child of five, six or seven might be bringing in 50 per cent of their own calories. If a snack requires processing, however, it will take more than goodwill for a child to produce it. As an example, children of Yucatec Maya farmers in the Puuc region of Mexico's Yucatán Peninsula can certainly cut and carry enough maize to feed themselves, and in little to no time, but they don't have the skill or the ability to soak, smash, grind, shape and cook the kernels into food. The move towards processed foods may produce more calories, but these new calories require different methods to extract them – and while children can help, anyone who has ever let a child stir a pot will know exactly how well that goes.

'Work', when we define it as the business of acquiring calories, takes up anywhere from almost no hours of a day to a maximum of about six in foraging groups – and this includes adolescents who must spend long hours with skilled adults beefing up their hunting and foraging skills. As survival becomes more dependent on a specific food-production technology, kids get roped in for more specific and defined tasks, and for longer. This intensification applies to the whole family: as the workloads of mothers increase, the workloads of their children are forced to keep pace. Grete Lillehammer, a pioneer in the study of children in the past, thought that the added demands of an agricultural lifestyle, particularly on mothers, were responsible for putting an even greater importance on the efforts of children in keeping a family afloat. Among the children of farmers, the number of hours worked is higher, between one to seven hours for boys and three to eight hours for girls. Even if they don't work terribly efficiently, children still contribute to the household energy budget with boring, unskilled bits of farming that 'even a child can do'. Only one way of living demands children work harder than farmers and that is making a living from herding animals. Communities that live by the goat (or cow, or sheep, or pig) have realised the unique potential of children; namely, how capable young persons are at staring blankly into a landscape while minding goats.*

When it comes to child's-play-that-is-actually-work we might do no better than to look to the fine art of playing with mud. Pottery itself predates our obsession with farmed foods – there are examples of clay figurines and bowls stretching back 20,000–30,000 years before the present. But an uptick in making things out of mud, attested by an uptick in broken bits of pots in archaeological finds, goes hand in hand with the storage revolution of the Neolithic age. Like any tool, this technology can either be produced well or ... less well. Pottery might be the perfect example to look at how children learn to be adults in societies where a sedentary lifestyle is starting to allow the accumulation of *things*. The potter or, perhaps more accurately, the family of potters, might have been an early semi-specialisation in much smaller concentrations of

---

* No one works harder than a pastoralist's children. Despite the extreme fetishisation of the animal-herding lifestyle (Marie Antoinette, ancient Greeks, famous historian Emmanuel Le Roy Ladurie), very few people seem to have put together the reason that all of these shepherds have time to compose sonnets, think deeply on the nature of the universe, *etc.* is because watching a bunch of ruminants all day is soul-crushingly dull.

humans well before the time of the first cities brought about the phenomenon of 'careers'.

Ethnographically, we can see that potters often involve their whole families in production, with tasks being doled out according to skill. Potter's children are necessarily involved, and their families invest heavily in the sort of social learning that will allow them to participate in the potter's world in future. For instance, in addition to providing the observational learning opportunities we expect for any child, at any time, a skilled adult might turn out a few pots for a child to actively practise decorating. A study of several hundred sub-par pots made in the American south-west around a thousand years ago by cultures ancestral to the Puebloans shows exactly how these trainee potters were given the chance to learn the craft by picking out the evidence of collaboration between competent potters and their not-so-competent colleagues: well-made vessels were given over to amateur decorators, partially complete decorations were finished off with rather less skill; sometimes these trainees were only allowed the simple job of colouring-in a pattern or a set area. Using clever measurements of fingerprints, or even the length of fingernail impressions, left in wet clay of Sinagua culture ceramics some 2,000–3,000 years ago in what is now Arizona, kid potters have been identified as responsible for some terrible pottery in the past, well before the invention of the kindergarten souvenir ashtray.*

---

* Note for younger readers: in the Bad Old Days, small children who were introduced to pottery-making at school were encouraged to make ashtrays due to their easy shape and, presumably, constantly smoking caregivers.

And, so, we see that children's contributions to society are a little easier to find when we live in a world of specialist crafts. The hyper-specialisation of work in urban economies means that playing at being a grown-up is going to have to fracture into just as many pieces as there are cogs in the means of production. A straightforward farming life is of course going to continue to exist in an urban economy; up until very, *very* recently most populations still made their living directly from the land, with varying degrees of engagement with urban life.* But, with new ways of producing capital, we get new ways for children to contribute, whether they like it or not.

One of the first written records we have of children put to work comes from the cloth-obsessed cultures of Mesopotamia, some 2,500 years ago. The natural resources of the 'Fertile Crescent' are on one hand, abundant – there is water, soil, plenty of people – and on the other hand, extremely limited: there is only water, soil and plenty of people. This is a tough platform to build an empire on unless you can leverage that tedious agricultural bounty into something that will get you what you want, which, in the case of Mesopotamia, was everything. By around 4,000 years ago the various polities of southern Iraq had hit on a winning formula. With the spare hands that all that agricultural bounty brought them, they could branch out, do something different, something that would add value to what they had and earn them a nice return. That added value would be the capital that a clever polity could leverage

---

* As enumerated by Johnny Cash, a twentieth-century American farmer might be pleased to own approximately five major non-mechanised farming implements, animal resources such as a large pig and a hunting dog, and a little less than a dollar (in dimes) and still be considered to be doing all right, for country trash.

into the ability to conquer enemies, raise temples and build an empire.[*]

The get-rich-quick scheme that applied most easily? A massive, state-sponsored garment industry. Cloth has an often-neglected role in the history of the world. For one thing, it doesn't last down the years, which makes it hard to notice in the archaeological record, and for another, cloth in our own time is extraordinarily cheap and easy to obtain. All of the critical symbolic business of dressing – showing group affiliation, status, idiosyncratic perceptions of harmonious colour – can be, in most places in the world, acquired for the price of a pizza. This shocking devaluation of a commodity that once ruled the world[†] was still a long, long way in the future for the Mesopotamians who put their excess population to work in textile factories. We know this because they wrote it down. Rather more specifically, they wrote down the important administrative aspects of the lucrative trade, including the rations and remuneration required for the weavers – children among them.

In the first cities of Mesopotamia, the temple was the only real employer outside of the home. The centralisation of resources and involvement in making things (that would be traded to acquire yet more things) gave temples a unique role in early society as the major engines of the economy. As the biggest employers in town, it was temples that managed the lives of the weavers. They were not necessarily

---

[*] Or, in my favourite phrasing, build 'a mound of your enemies to touch the sky'.

[†] Even I can't think of a way to shoehorn into this discussion the fascinating history of Chinese silk, or the market systems of the medieval Eurasian cloth trade, but it really did shape the modern world.

completely free – some were captives from military adventures, while others were unfortunates and the destitute who were obliged to look to the temple to take them in. Temple complexes seem to have been the social welfare agencies of their times, taking in far more women and children than men and not only that, accommodating widows, the infirm, the illegitimate and orphans. We know from the set of texts sometimes referred to as the *'gemé-dumu'* texts, or 'women and children' texts, that the workers in these places were regularly fed. Of course, feeding an orphan 20 quarts (more or less) of barley a month is a small price to pay for getting a lifetime of service making the cloth that drives your economy.

But how are we to trace the part children play in the economies of cities, states and empires that *don't* merit a mention in the tabulations of worker productivity that the Mesopotamians cared about? There is the small chance that the written word itself can be co-opted to give us the voice of children directly. Small, because we have only a tiny, tiny proportion of the total texts ever produced in all of written history that have survived to this day and because those texts that do survive tend to survive for a reason – they are canonical, important, weighty* texts that are deliberately preserved. It really is only in exceptional circumstances that we get to see the scribbles of actual children and have the additional clues to help us identify them as such.

One such exception is provided by the eruption of Mount Vesuvius in 79AD, which covered the towns of Pompeii and Herculaneum in a thick layer of burning ash so quickly that the entire townscapes were preserved as though overcome by a particularly corrosive, fatal kind of

---

* This is quite literally true of any writing on clay tablets.

amber. Along with the bodies and the remains of last meals, graffiti of varying degrees of wit was preserved, and has remained etched on walls for thousands of years; certainly well past the potential usefulness of knowing that an imperial finance officer declared the food at the house of Paquius Proculus to be 'poison'.*

Alongside such valuable cultural information there are around 160 pieces of graffiti at Pompeii that bear the hallmarks of potentially being made by children. The reckoning is that if a graffito is found low down on a wall, it could have been scratched there by a small-statured child; similarly, if an image or word looks rough or poorly spelt, we might think it made by a child who is still learning the relative size of a boar versus a dog, or the number of 'm's in the girl's name 'Mummia'. These aren't actually hard-and-fast proof that children made the inscriptions, of course, and many scholars disagree that poorly drawn boars should be immediately identified as the work of children – adults can be terrible at drawing animals too. Others have argued that stylistic elements of children's drawings are pretty universal – the giant heads and after-thought bodies, the bizarre super-position of objects that reflect what the child *knows* to be there but certainly not what would be visible – make a strong argument for being able to see the direct actions of children in the past.

One of the main themes in the 'art' attributed to Pompeii's children is animals. Some of these animals are birds or other decorative 'pets', but quite a few are actually

---

* Though writing your review of a meal *on* the building you actually had it in is pretty expedient, one assumes the insulted host would have been out in the morning to chip off the zero-star review from the hall column his illustrious guest had scratched it on. Had there been a morning.

useful domesticates like pigs, sheep and goats. As we discussed previously with regard to the children of pastoralists, one of the easiest jobs you can set children to do is caring for livestock. It may be that these children, 2,000 years ago, were idling scratching out images of their domestic chores.

If you want to look for work from the hands of children themselves, we have to look again to special circumstances that have preserved the not-quite-adept work of little hands. In the case of the children of the New Kingdom of Egypt some 3,000 years ago, we have some small clues as to what they may have been up to solely because of the discovery of a vast trove of scribbled 'ostraca' at the Theban site of Deir el-Medina. The scrap paper of the day, ostraca are bits of limestone or broken pot scribbled on with little dedications, images or other ephemera. Quite a lot of the ostraca at Medina appear to be created in the process of learning to draw, or even to spell out cartouches and practise handwriting. Figurative art had an important role in dynastic Egyptian art, of course, so mastering the skill of representing the world might lead to a career in the workshops at the Valley of the Kings decorating the temples and graves of the elite. But, as the variable quality of the images on these ostraca suggest, illustration is a skill that requires work.

The corpus of discoveries includes wiggly, uneven strokes making up pharaohs with unlikely facial proportions alongside skilled practice sketches and everything in between. A monkey climbing a tree is not too shabby an effort until you realise the monkey is covered in spots and therefore a monkey totally unknown to science, suffering from a deeply unfortunate skin condition, or not a monkey at all but a very compact leopard. Of course, you could also look at that unlikely monkey and see behind it the hand of

Figure 19.1. *Date-picking baboon with unlikely spots, 1315–1081* BC, *on a painted ostracon held at Los Angeles County Museum of Art (gift of Robert Miller and Marilyn Miller Deluca; M.80.199.50).*

a young artist who has never seen such a creature up close, but dreams nonetheless of expanding their horizons along with their artistic repertoire. But what we do not know for sure is whether these trainee sketchers are children or merely bad artists, and if they were children, how experienced or how old they were when they worked at their craft.

Another way to look for children at work in the ancient world is to, quite literally, look for them. We can search the friezes, statuary, painted vases and engraved metalware of the ancient world for depictions of children at work, but there are very, very few instances where you might reasonably call what children are pictured doing 'work'.

From ancient Greece there is a depiction of a girl watching her mother weave on one of the plaques from the famed Athenian Acropolis, and a fragment of another plaque that has a female child messing about with balls of wool while her mother weaves; other images of women weaving include what may be teenage or younger adolescent girls with unbound hair. Textile production in ancient Greece was a critical industry just like it was in Mesopotamia, and rather notably the sole expected employment of a certain class of woman, so it makes sense that of the very, very few depictions of female children we see, they would be remembered in the context of rich ladies.

One spectacular site, however, offers a very unique perspective on children in ancient Greece. The island of Santorini is, today, a mangled volcanic rim around a stunningly picturesque caldera,* mostly known for its whitewashed villages set into the dramatic cliff face of the caldera. Several thousand years ago, Santorini was Thera: a nice island with good sea-connections and a reasonably big Minoan-flavoured settlement at the site of Akrotiri. Around 1600 BC, however, the whole thing went bang – quite literally. The volcano at the heart of Santorini began to erupt, churning out ash, earthquakes and little tsunamis. Extremely fortunately for the people of Akrotiri, however, the eruption took some time to build up

---

* Picturesque, if you don't mind the innumerable cruise ships stacked up in the sparkling Aegean water like cars in the parking lot in the aftermath of a football game. Santorini is extraordinarily beautiful, with an entirely under-visited hinterland full of vineyards and beaches, but the sheer volume of day-trippers to the cliffside villages at the top of the caldera prompts the kind of existential despair that can only be dealt with by climbing on a roof with a good amount of gin and refusing to come down until they have left. This is also a good way to see the famed sunsets of Oia.

steam – or magma, rather – and they seem to have been able to flee the town with most of their valuables; there aren't many high-value metal finds left at the site, and (happily) no pyroclastic moulds of human bodies like at Pompeii or Herculaneum. Shortly after they left, there was an eruption so massive that it deposited nearly 60m of ash on the island, leaving a blanket of choking dust and rock in its wake. The eruption was sufficiently massive that it may have sent earthquakes and tsunamis that wreaked havoc on the Minoan civilisation of Crete, and was noted as far away possibly as Egypt where there is a storm and a 'darkness that covers the two lands' and even China, where texts talk of a year of frost, famine and dim sun at the end of the Xia dynasty.

Akrotiri was buried in pumice and ash, up to the second storey in some cases. This left the actual buildings intact, with all of the quotidian materials from the daily life of Therans 3,000 years ago preserved, from pots to the empty voids left by wooden furniture and, of relevance here, frescoes. The painted decoration of the house walls of Akrotiri give us an unparalleled insight into the lives of children, representing them going about their days in ways we might guess are fairly standard to ways that might be just a little fanciful. There is a lovely image, for instance, of a girl picking crocus stems and putting them in a basket, in the house of Xeste 3; the mundanity of such a task is rendered a bit questionable by the following panel, which shows a blue monkey taking the stems and handing them to a woman on a throne, while a leashed griffin hangs out in a window. But then, these are not ordinary girls. With their ritually shaved heads, fancy clothing and the red dots of their carnelian jewellery picked out in paint, we are seeing a distinct Minoan career option for children: *acolyte*. Just as in Mesopotamia a millennium earlier, special roles for children can be found in the number one specialist employer of children, the temple.

It is difficult to work out what daily life would have been for the vast majority of children when the imagery we have comes to us from the top of the social heap. Art, at least the big impressive bits of it that have survived thousands of years, is produced by elites and features elites. The only reason we ever see other sorts of people is when they intrude on some elite scene in the course of their work – holding the king's lions, nursing the queen's babies. But where art starts to touch on the mundane, we see little clues as to the role of children in everyday life. In the endless iconography of the ancient Egyptian world, for instance, there are plenty of children doing chores – they light ovens, they harvest grain, they carry water, they carry other children, they even make pottery. The hoard of ostraca from Deir el-Medina suggest a wealth of employment opportunities in ancient Egypt undreamt of in the monolithic economies of early village life: along with the various agricultural labour that could have been carried out anytime in the past 10,000 years, children are depicted as messengers, carrying bureaucracy from place to place. Ostraca also depict the rather specific job of date-retrieving-baboon-trainer, a niche and perhaps low-status occupation involving getting 20–30kg of large-toothed monkey to go up a tree and pick dates that seems to have been the province of boys and foreigners.*

A final source to consider for information on children's work comes from the bodies of children themselves. As we saw in earlier chapters, what we do with our skeletons

---

* This is a rather unfair situation to contrast with that of the acolytes of Akrotiri, where monkey-adjacent employment seems to be of much higher status.

leaves behind traces on the bones themselves. In the case of trying to trace activity in children by looking at the workload on their bones, however, we are really stuck. The principle that we get bigger bones to hold bigger muscles to carry heavier loads still holds; but our skeletal system has to balance the demands of growth against the need for structural stability. Changes in children's bones caused by repetitive actions, heavy loads, or even overuse and injury are obscured by the fact that children's bones are unfinished, growing and constantly in a state of flux. However, in the most extreme cases, even malleable, still-growing children's bones can get worn down by work.

We know this because we can see it, in evidence like the 2,000-year-old bodies from the mining community of Hallstatt, Austria. Hallstatt was a salt mine, off and on, for thousands of years, and salt mining is not a pleasant activity. It involves repetitive, awkward motions – and a lot of them. The adults at Hallstatt have all the hallmarks of such strenuous activity – the joint breakdowns, the arthritis, even little splinters of bone where they've sheared off bits through dint of the force their muscles are applying. And shockingly, of the 15 young people's skeletons that could be studied so far, more than a third showed some sign of arthritic degeneration. Backs, elbows and ankles were all on their way to being knackered in children from eight years old onwards. Researchers suggest that to cause such destruction in their spines these kids may have been hauling heavy blocks of salt either directly on their heads or by using a head-strap. The children of Hallstatt clearly did not have easy lives. Work was actively dangerous, and their skeletons testify to not only the demands placed on them in life but also the sad fact that they did not survive to adulthood.

The workers in the salt mines of Hallstatt and the weaving orphans of Mesopotamia had one thing in

common – and that was their position down at the bottom
of the hierarchy of urban society. The ability to hold your
place in the snakes-and-ladders social whirl of the ancient
world relies on the ability of a child's family to teach them
the skills they will need to do the sorts of jobs they were
born to do. As I said at the start, for the vast majority of the
human experience and the vast majority of jobs, the
training process is in house (even when it is god's house).
But what happens when the skills you need to learn to, say,
build a monumental temple are so complex and take such
time to acquire that you need to start looking for outside
help to learn them? The final aspect of childhood to
consider is exactly this pyramid scheme of teaching teachers
to teach new teachers, which sees the expansion of formal
teaching into ever greater areas of our lives and has perhaps
wrought some of the most significant changes in how we
expect our children to spend their childhood.

## CHAPTER TWENTY

# How Many Miles to Babylon?: A Very Human Childhood

How many miles to Babylon?
Three score miles and ten.
Can I get there by candle-light? Yes, and back again.

A t last we come to the one final strategy for childhood that only our species has made, an investment strategy that humans, and only humans, have come up with. As we've seen, you can't make any sort of baby without some embodied, physical capital; and social capital is a feature of most animal life – even pigs play and crows have friends. But only we have found a way to leverage material capital into better chances for our children. Once that very first person figured out how to turn things into influence, we mutated into a unique creature with a third way to build our children: building up resources that can be exchanged for advantage. In so doing, we have riven our kids into haves and have-nots, and given our species not just one but many types of childhood. Our long childhoods are no longer equal, with the children *here* having just as much chance to prepare for successful adulthood *there*.

It's not too hard to see this in practice. We saw in the previous chapter that our species has also, rather stupidly, invented work and our children are expected to contribute in all sorts of ways to their society. In many of the cases we saw in the previous chapter, work is where the investment in our offspring starts to turn around and pay dividends. You might think it's about time, given the drawn-out nature of our adolescence and the massive investment we make in keeping kids alive and fed; after all, adolescence is when primates start to sort out their adult social roles and chasing a date-picking-monkey around a palm orchard is a valuable, if weirdly specific, contribution to society. Kids have to grow up some time and make the transition from being an investment to an investor. But if you have enough material capital to keep feeding that kid and keep it playing and learning, well then, you don't have to send your kids down the mines. You can take that capital and buy your child a better life. And *that* is the world we live in today.

Success in life depends on the investments families are able to make in their offspring. We saw that in its most basic form in the early chapters of this book and throughout our species' evolutionary history – in the ability to physically provide the nutrition to conceive, carry and raise a child. It also depends on a type of social investment, the relationships and knowledge that secure a child's place in the community and allow them to negotiate group living. For humans in small, mobile groups, this might take the form of teaching the stories of land, kinship or cosmology needed to get along, and taking the time and effort to ensure children learn the things they need to learn. For those experimenting with settled life in a fairly undifferentiated social terrain, where everyone is more or less socially equal,* kids get the same kind of investment. It is when humans start gathering in big groups, and those groups start to organise into tiers

---

* 'Egalitarian' is a ridiculous word to use for the past, when we can clearly see that social roles were different for different age and gender groups. Really, just the fact that the past has humans in it should be a giveaway that 'equal' does not always apply.

and hierarchies, that we see the first major difference in *material* capital invested in children – and we see it first in the dead, in the material resources their families gave them to take to the grave.

As we've discussed, the children of our long, free-range past were loved just as much as we love our children today, and when they died they were often buried with impressive adornments. The famous 40,000-year-old double burial of two infants in the Grotta dei Fanciulli still contained rows and rows of decorative shells, all out the patterns in which they were sewn on the infants' final clothing. Other substances like ochre were often used to decorate a burial, or animal parts; there is a star-shaped interment of big fallow deer antlers in the grave of several infants discovered in the 15,000-year-old burial ground in Taforalt cave. But while decoration and personal adornment was common, reading special status into the artefacts we find is difficult. Only one Upper Paleolithic burial might be said to have an object in it that was a *real* thing, of actual use, twin knives buried in the hands of a young child.

However, as we get to a world of villages and settlements, we find children are buried with things. Things they would have had in life – beads and decoration, dustings of ochre, and perhaps a few blades or tools they would have used. However, while there might be subtle differences between who is buried how, and what they are buried with, the things we find in burials seem to suit social roles – adult versus child, male versus female, shaman versus the spiritually underwhelming. There is little differentiation between this child and that, or this woman and that one, just a few bits of knife or bead.

All of the new technologies of the first farming villages, however, cannot possibly prepare us for the mind-blowing explosion of *things* that appear as we start to live in the more populous huddles of towns and cities. There are metal things, written things, ground things, carved things, ceramic things, carved things that look like metal things and ceramic things that look like carved things – the urban past is a junkheap of consumer items and quite a lot of it gets taken to the grave. We care about this proliferation of objects because they stand in for this new kind of invest-ment, one that has been difficult to trace in the archaeological record in less clearly stratified societies: material capital. Every object in every grave since the beginning of time has, of course, some capacity to be considered as material capital. The shells strung together in the Grotta dei Fanciulli double burial obviously required investment in terms of labour and materials. But lacking a way to leverage or exchange these for other items of greater labour or material cost, because almost everyone can produce the same beads from the same materials, there is not much of an invest-ment lost when the beads go into the ground.

However, when only some people make beads, and only some people have access to the raw material for beads, the economic meaning of those beads changes. And this is what happens to every class of material object as the means of production and the mechanisms of exchange stretch from their community roots into what we understand today as a market economy, where things suddenly have prices that are detached from the context in which they were acquired – where who made the gift matters less than the gift itself. Now that everything has an exchange value – and there are so many things – when we see them distributed in the graves of the dead, we can start to make some very hard-headed assumptions about the ability of those dead to enact their will on the material world. Things, and the ability to leverage them, become a critical part of survival in an urban world, and therefore what tells us the most about the evolution of a new form of childhood lies in the differences we can see in *things*. Cold, hard, *things* are not the embodied capital that parents invest in growing their children, or teaching them to walk; they are not the social capital that manages the complex network of human relationships needed for a child to thrive in society; they are steps on a ladder, up or down. They are money.

We can see this clearly in the archaeological record. As denser human societies start to organise themselves in hierarchies, graves get *fancy*. Yes, there were some fancy graves before.* But we do not need a material hierarchy to explain what could be shamanistic or otherwise cultic burials; and indeed, this is the explanation for most of the 'special' burials found in the era before life gets stratified. But somewhere along the way, things start to accrue in graves. For adults, we can see that the things that go into the ground when someone dies are reflective of their own personal items and accomplishments. The classic example here being the 'warrior' grave, where weaponry alongside a dead body† is interpreted as clearly marking the fighting prowess of the deceased. Metal is a stupid thing to throw away, however. It's expensive to extract and forge; particularly in those cultures that favour metal for their

---

* My personal favourite is a disabled elderly woman buried at Hilazon Tachtit 12,000 years ago with literally thousands of tortoise shells. You do have to wonder at the person who looked around at the first few hundred and decided, no, this is not enough tortoise shells.

† Not a dead *man*, a dead *body*. There are sure to be plenty of red faces wherever archaeologists go when they die, blushing for shame at having assumed weapon + skeleton = male only to be corrected by a helpful bioarchaeologist who actually *looks* at the skeletons. Rather miraculous how many female warriors started to turn up once we learned how to do ancient DNA.

tools.* So when you start burying high-value goods with your dead, it must be for a reason – either their prestige or yours. This is all well and good, because we can understand that an adult has plenty of time to earn whatever prestige – and metal spear – they take to the grave with them. But what does it mean when we start to see, for instance, the weapons of war, vast piles of them all made of expensive metal – in the grave of a child?

This is a question I have had to ask myself – and it is a fascinating one. In 2015 I joined an excavation project at the site of Başur Höyük in Turkey's south-eastern province of Siirt – on the sort of northern fringes of the Tigris river in the cultural area we call Mesopotamia – signing on with Dr Haluk Sağlamtimur of Ege University's team as they excavated an Early Bronze Age cemetery. The team had uncovered tomb after tomb full of elaborate pottery, complicated beading, and piles and literal *actual* piles of metal objects – all dating back to the very earliest days of metal use in the region 5,000 years ago. Each tomb had a few occupants and then, set *outside* the big stone borders of the grave, a few more – also buried with a flourish of pots and precious objects. And in the biggest, and deepest grave – perhaps the earliest one – it became clear that not only had the pots and metal objects outside the grave been carefully laid down in honour of the tomb burial, but the actual people had as well.

Eight young people, aged from about ten to their early twenties, were piled on top of each other at the foot of the main tomb burial, just outside the stone walls that guarded the tomb. Painstaking examination of the cracked bits of skull and shattered limbs revealed that these youngsters had

---

* Well, if you should happen to construct your urban polity in a region where the metals are mostly soft like gold, as in Mesoamerica, you're going to stick with stone, aren't you.

not had a good end. At least two of them were clearly violently killed. So here, at the beginning of the urban revolution, just outside what we would call Mesopotamia, was something astonishing: graves full of goods – ceramic, metal and *human*.

So, who did these young people's deaths honour? Who was inside that tomb? The human remains were very carefully collected, despite being in medium-to-rubbish condition, and when I came face to face with these mysterious grave-builders in the lab on site, it was my job to figure out who on earth they *were* to merit such fabulous wealth.[*] The bones had largely fragmented[†] so my task was to identify the individual pieces – a fragment of the chunky bone that surrounds the ear, a few teeth still clinging to a jaw – and then decide who they belonged to. For adults, this is never an easy task, and you end up counting how many of each body part you have in order to make a reasoned guess; *i.e.* three left legs mean you have at least three individuals. But for the still-growing bodies of children, we have the known trajectories of growth that we saw in previous chapters to narrow down exactly what age the body is. By using information from the bones (and the teeth; mostly the teeth) – how many of each part there were and the stage of development – it became clear that the main

---

[*] It was also my job to pay for the supply of gin needed to keep my English assistant on board throughout four months of 5 a.m. baby-goat-based breakfasts (buryan – a local delicacy) and 50°C heat, not to mention the ice-cream sandwich provision for the students who digitised some thousands of skeletal parts. I also had to look out for the rogue turkey who kept trying to insert himself in the finds analyses process, beak first. Archaeological work requires a certain flexibility.

[†] If you would like your remains to *remain*, avoid being buried on limestone. Top tip.

honourees of what might be the earliest example of human sacrifice on the planet were two 12-year-old children.

It is an extraordinary thing to imagine that two children could command such respect in death that it could be accounted in actual human lives. This is another world entirely from some shell beads, or your personal knives. There is a fundamental difference in worlds where status comes through actions in life and worlds where status is a function of birth. The difference between a culture of earned status (the warrior with her spear) and ascribed status (the child with weapons they cannot hope to wield) is immense, and it is the largest chasm between the childhoods of our long past and the childhoods most of us experience today. It is the evidence of capital that is passed down, inherited. It is the sign that in this society, not all children are equal.

In dense human societies where some are more equal than others, social status becomes a thing you can invest. Your social capital – belonging to the correct family or lineage, this class or that profession – becomes transmissible in actual material form; in metal spearheads or whatever else your culture values. Whereas life on the hoof or in small villages keeps hierarchies mostly flat – and social capital only extends as far as you can teach your children how to live in their worlds – suddenly, the urbanite has a store of rank and privilege (or lack thereof) that is going to go down the generations. And in that tomb in the alternately snow-and-sun-scoured corner of Turkey, we see some of the first indications that this inequality in social capital is going to be buried hand-in-hand with the trappings of real, material goods – be they fancy metal spearheads or actual human lives. In the mirrored fates of the bodies inside and outside the tomb, we see the starkest example possible of what it means to be a child *with* and a child *without.* This is the story that writes our modern world and we can see it throughout the archaeological record in the bodies of the dead.

The fact that most of what we know about childhood in the several thousand years before the modern era comes from the bodies and burials of the dead is a stark one. A question I am often asked is how it feels to uncover the bodies of children. There is no simple answer and I imagine that my attempts end up being much the same way as anyone who deals with the darker side of human experience: you do it because it's important, and because some questions cannot be answered without looking into the most unpleasant parts of life. Most of us who dig bones were trained well before we had families of our own, and it may be that the callousness of youth is a useful asset in dealing with the dead. But I think there is more to it than that – I think the people who dig bones are actually often the *most* invested in the stories of the individual lives each skeleton represents. The profession of archaeological diggers of the dead now skews heavily female, which may not be unrelated to the fact that the records of life locked into teeth and bone are some of the only ways we have to access the lives of the women and children who are commonly shut out of written history. But the real answer is yes, even professional archaeologists can be affected by the pathos of the little lives that never were. It can be crushingly sad, and this is the problem with digging up the lives of children – so many of them never had a chance. But if we want to move beyond a shallow understanding of the human past, written by history's winners, dig we must.

This is the flipside of children in massive tombs with piles of hoarded wealth to accompany them to the next world: children whose capital did not extend to earning them the elaborate burial rituals of the elite. Deprived not only of social capital but the investment in their physical bodies too, their small skeletons show the real consequences of stratification: over and over again, they died for lack of capital. The rise of hierarchical urban living is the rise of malnutrition; the rise of unequal access to resources that see some families able to invest in the physical growth and health of their child, and some families powerless to protect them through networks built on social capital. Malnutrition, particularly when it comes to specific vitamins and minerals, is something that has lasting effects on the skeleton, so it is not too surprising that the point of human history where we start to see the telltale signs of diseases like scurvy and rickets is the same one we see the rise of people living in societies where some have and some do not.

The earliest case of scurvy in ancient Egypt was discovered in a village at the very southern edge of the Egyptian cultural sphere, in a cemetery dating from almost 5,000 years ago, and it was an isolated occurrence. By the time the Romans were messing around with Egypt as part of their empire, 3,000 years later, we see in the much-abused bodies buried at Dakhleh Oasis that scurvy might have been present in as many as a quarter of the children buried. Some 60–80 per cent of children buried at Dakhleh have the characteristic under-eye lesions scurvy causes. We can see that these diseases of want are not distributed equally, however. In the Harappan sites of the Indus Valley, another early urban society, there are children buried with no evidence of malnutrition in one burial ground while in another, a third of children had evidence of scurvy. While scurvy and rickets were never as common in the Americas as they were in the Old World, there are other measures of childhood health that show a dramatic increase with socially stratified urban living. Lines on teeth caused by episodes of childhood growth disruption become a key piece of evidence, showing that children in denser urban environments were far more likely to experience some sort of interruption to their growth. This is true in cultures as far flung as from the Inca in fifteenth-century Peru to the children buried in the Shang dynasty cemeteries of Yinxu in the twelfth century BC. And this is hardly an archaeology-only problem; under-nutrition still contributes to *one-third* of the more than 8 million children who die every year.

But, and here's the thing. What if you could buffer the risk of an unequal world? Because that is what humans do. We saw previously how drawing out the period of investment is a valid strategy for growing in risky environments; if you can't eat enough to grow now, hopefully you can make it up over the long run. Thinking about the new, and very human, invention of material capital, how do you make sure that your kid has enough to succeed? Well, we do what all animals do, which is give ourselves longer to make sure we get it right. We take that long, drawn-out period of investment in your child and make it longer, giving our kids a chance to gird themselves for the battles ahead. So, what if you could take all that capital you have for your kids – the food, the friends, even the cash – and just keep doling it out? What would you do with that extra time? And it turns out this is a very important question, because the structure of the societies we live in now lean very heavily on the answer. We (mostly) don't leap into reproduction the second it becomes a biological possibility. We take another decade, possibly two, to remain operating as a net loss. Our cultures have built structures to make sure we make the most of our adolescence; we tell our children when they can marry and when they are adults – and usually this has sweet nothing to do with biology, but everything to do with when we think children should transition from being invested in into investors themselves.

We offer a final category of training to our children, one
that the temple acolytes and ostraca-artists would recognise.
We teach them how to reproduce their culture and to
maintain their roles in it; we invest every type of resource
we have in this final, most critical endeavour. In every
human culture, we know things that have to be taught. It is
a difficult business, knowing things. Imagine the time it
takes to learn the songlines of indigenous Australia; to
memorise the Qur'an; to learn where the mammoths go.
Now think of the advantage the child learner has whose
only job it is to learn. In societies around the world, we have
indeed thought about this, and thought it was a very good
idea. We rattled our resources around, spread them out and
decided to give (some) of our children the chance to be
children – absolute resource sinks – for a long, long time.

Leveraging your capital into a longer period of
dependence so your children can learn things is hardly
new. It is clear that parents invested immensely in every
possible way in their children in the past, even if the
children themselves did not appreciate it. The invention of
correspondence some 5,000 years ago gives us a fairly
startling insight into the inner workings of these
relationships, even ones long, long lost to time. For instance,
it is not difficult to imagine a reworking of the Old
Babylonian period letter from one Iddin-Sîn to his mother
Zinû into a set of testy text messages in the modern day:
'The son of Adad-iddinnam, whose father is a servant of
my father, has two new garments to wear, but you keep
getting upset over just one garment for me. While you gave
birth to me, his mother got him by adoption, but you do
not love me in the way his mother loves him.'[*]

---

[*] See reference in Veenhof, 2005.

We can immediately understand that Iddin-Sîn is somewhere far from home, looking to his mother to outfit him to succeed in the world – and getting peevish as hell when he perceives her support to fall short. Here is the latest and most lasting trace of childhood. We encounter, at last, the child who is on the border of adulthood, trained up in the art of writing, but still bothering his mother about laundry. Iddin-Sîn tells us about the last step on the shaky ladder to growing up human: outsourcing. Just like the adolescent chimpanzees finally getting a chance to hang with the big boys, we have a phase where society is going to do the job of raising us. What chimps don't have, however, is the material capital that can buy them (or cost them) their chance to succeed in that final push towards adulthood.

Humans, however, do, and we use it as best we can. We saw previously that the vast majority of 'professions' in the past were family affairs. Crafts were passed down within family units, something we see attested by the very fact that it's only the exceptions that merit a mention in really ancient texts. The idea of investing cold hard cash in your kids leaving the home to learn something was sufficiently unorthodox for the Babylonians 3,000 years ago to need to set in hard clay the terms of engagement. Apprentices weren't even necessarily better off than those keeping it in the family. Quite a few were slaves and the money exchanged for their training wasn't very impressive: one apprentice mouse-catcher is to be compensated the not-terribly-princely sum of 50 mice a year – and this was to be apprenticed to the mouse-catcher of the king.* Many of them might have been boys sent to learn a trade when their own fathers were dead,

---

* Contract penalties could be exacting: should the king's mouse-catcher have failed to train young Šamaš in 549 BC, there would have been a fine of a *thousand* mice.

paying a very small fee to learn woodworking, sandal-making, mouse-catching or whatever it might be.

It is all very well and good to be trained to a profession, a career, some way of making a living. But what of training for the *sake* of training? Education, for the hell of it? Well, for one thing, learning is never really extraneous. Our Palaeolithic ancestors – and our hominin ancestors before them, and the apes and monkeys and tree-shrew-like-things before them, all *learned*. It may not be easy in the eyeball-sized brains of tarsiers to work out what, precisely, they learned in the course of their lifetimes, but by the time we get to the complex societies of monkeys and apes, we can clearly see the pattern of observation and imitation that turns a dropped sweet potato into a saltwater snack. Learning how to salt a sweet potato from macaque peers and relatives is no less important than learning the cosmology that explains your existence – and the workings of the world you live in – both are adaptive strategies that transmit knowledge *culturally*, through socialisation. Only the latter, however, gets labelled ontology and, as soon as it is captured and forced to ossify in written form, elevated to the status of *learning*, something we would do well to remember as we look at the types of training we invest in for our children. As persons inhabiting a world dominated by written language and the types of learning that go with it, we have a habit of discounting the immense investment societies make in teaching things to their children because we do not recognise the form they are taught in. We dismiss the type of learning that can be done passively, through observation and imitation, or through guided experience and demand to be shown a didactic education.[*]

---

[*] Because, heaven forfend, you don't know the kings and queens of England. I mean, that should definitely be taking up space in your brain. Not like you could just look it up somewhere.

Education, we imagine, is removed from immediate practical application. The invention of 'school' as a place for producing well-rounded knowledge is held up as a pinnacle of civilised achievement – learning for the sake of learning. But it is absolutely clear that, from the beginning, even the most formal, esoteric instruction (looking at you, ancient Greece), had real-word intentions. The goal of education, or schooling, or the sort of formal pedantry that we have become immured to in our modern societies, is always, *always*, to produce a set of socially learned skills that allow the pupil to function in society. In exactly the same way that you need to learn how to get along with others or your small group of hunters is going to rapidly become even smaller, in dense social environments you are going to have to find ways to signal your suitability to contribute to the group – except instead of knowing the lengthy verses of a saga relating to the seasons or landscape, you're going to want a handful of *bon mots* on *Aeneid* up your sleeve.*

The earliest incarnations of this new form of specialist training is in a skill that only bureaucrats could love – writing. The stock symbols of market exchange arose first, some time before 6,000 years ago, because of the absolute need to communicate that *this* jar has oil in it and *this* shipment contains two jars and two bundles of cloth, when those doing the exchanging couldn't speak directly. Tokens representing goods and symbols representing sellers were etched into the omnipresent Mesopotamian mud either sealed over the top of containers or in little sealed clay envelopes so that their contents could be guaranteed undisturbed. Pretty soon counting tokens

---

* In the UK, this actually appears to be the sole criteria required to enter politics from a certain background.

becomes counting notches that mean the same thing as tokens, and before you know it, it's the third millennium and your symbols have become syllables and your mud has become literature. Even when text is at its most tedious, it still requires considerable learning to create. The symbols that move the sound of the spoken word into a visual representation have to be learned, as do the mechanical and motor skills of making both the writing and the writing materials. This, it seems, was every bit as challenging to do in Sumerian cuneiform as it is for children battling Latinate alphabets or Chinese logograms today.

From the city of Nippur, which was a reasonably big deal in Mesopotamia 4,000 years ago, a poetic description of the trials and tribulations of a child who attended the *edubba*, the tablet-house or school, is recorded in a text called 'Schooldays' from around 1800 BC. The boy – and only boys were trained in the schools, despite the fact that the patron saint of scribes in Mesopotamia was the goddess Nisaba and the first named author in all of Mesopotamian history was a priestess of Ur* – recounts his miserable day. After he gets his packed lunch from his mother and sets off for school, everything seems to go wrong: the hall monitor beats him for being in the wrong place, another teacher beats him for tying his shirt up wrong, the Sumerian

---

* Enheduanna was the daughter, probably, of Sargon of Akkad, and priestess of Innana. Her passionate hymns to the goddess ('You are known by Your massacring (their people), You are known by Your devouring (their) dead like a dog') and lament on a period of exile seem to have struck a chord that preserved her legacy for over 4,000 years, which is just a little bit like having Alanis Morissette's *Jagged Little Pill* be the only recording that survives until 6000 AD. If Alanis Morissette were a princess.

teacher beats him for speaking Akkadian* and his cuneiform
tutor beats him for just being terrible at cuneiform.
Nonetheless, the text dreams of the glory of being
eventually acclaimed a scribe and understandably so: boys
trained in such schools could go on to profitable
employment, not just because they could write – even
women could do that – but because they had been through
a school that inculcated sets of values – Sumerian, tying
your shirt straight – that would give them the kind of social
capital that can be traded for material capital, aka money
and influence.

   School, of course, has always been conceived of in exactly
the same way that you remember it. Not the classrooms
particularly† or even the manner of instruction, but the
*experience* of formal schooling appear almost unchanged for
millennia. In 424 BC, the playwright Aristophanes produced
*The Clouds*, a reasonably unsuccessful comedy that took
nearly all of its humour from the over-the-top depiction of
mad school master Socrates and the whimpering intellectual
detritus of his students. The voice of tradition pipes up in
the middle of the play to describe schooling of old, before it
was dumbed-down by namby-pamby (literal) sophistry, a
place where 'no one should hear the voice of a boy uttering

---

* The spoken language at this time *was* Akkadian – but scribes were
taught to read and write in Sumerian, which hadn't been commonly
spoken for hundreds of years – just as many thousands of years later
schoolchildren would be subjected to the extraneous cases of
doornail-dead Latin.
† Classrooms have varied impressively, from 'Teacher's rock'
('Daskalopetra'), an open-air, granite-floored auditorium on the
island of Chios from which I am finishing this manuscript and
where almost certainly Homer did not teach, to the open-air,
granite-floored auditoria of the most expensive 'forest schools' of the
modern educational movement.

a syllable; and next, that those from the same quarter of the town should march in good order through the streets to the school of the Harp-master, naked, and in a body, even if it were to snow as thick as meal'.*

The idea of a school as a training ground for a particular class of occupation – and a particular class of people who will occupy those occupations – seems baked into the origin of the institution. From the scribes of Babylon onwards, school has held a particular place in the production – and reproduction – of a set of cultural and social norms that cover what one ought to know, if one is going to grow up to do certain kinds of jobs. Investing time and money in sending children to school is very much taking the ability of capital to leverage social relations, and for the several thousands of years that we have been doing it, we have been slowly building up the capacity for that investment. If it used to take a decade to train a scribe, now it takes three to train an Assyriologist who will write about it. How did we come to the conclusion that we need all of that extra time, putting off the end of childhood and pouring more and more investment into the already biologically stretched adolescence of our species? And what does it mean for the shape of our childhoods?

---

* Aristophanes. 1901 translation. *Clouds.* This is subsequently rather undermined by the rest of the tribulations of a righteous schooling including a great deal of pederasty, which goes to show that not only is the 'walked to school every day in the snow, uphill both ways' trope of great antiquity, so are the majority of jokes it is possible to make about schooling.

## CHAPTER TWENTY-ONE

# Thursday's Child: Far to Go

Monday's child is fair of face,
Tuesday's child is full of grace.
Wednesday's child is full of woe,
Thursday's child has far to go.
Friday's child is loving and giving,
Saturday's child works hard for a living.
And the child born on the Sabbath day
Is bonny and blithe, good and gay.

We are not the only species that raises young. But we may be the only species that has ever tried to raise its young as intensely, for as long and using as many resources as we do. The heart of this book is a throwaway comment I've made multiple times over the years to explain why I have gone off to live in a room with an unreasonable number of chandeliers in Egypt for a year;* or a tent in Central Anatolia for months at a time; or really any of a variety of other questionable life choices I've made instead of doing something sensible like getting a job and starting a family. I have been, as I have informed many a parent, grandparent, and student loan officer, enjoying an urban childhood: 10 or 15 extra years where I am a net loss to society but reasonably fun at parties. Whether this is humour is debatable, but the reality is not entirely facetious. My life is shaped the way it is because someone gave me the chance to keep training, keep learning; the privileges I have are the result of investments I didn't make.

---

* Two is an unreasonable number of chandeliers for a bedroom.

Species by species, we all come up with our own
trajectory, the perfect solution to how long to spend in the
womb, at the breast or just knocking around annoying our
parents. The rate at which we do the growing that we do
is varied – and variable – and is decided by our investment
strategy at each life-history stage. Take pregnancy, for
instance. Some animals, like the year-and-a-quarter cooker
giraffe, do most of their life prep in the womb, so as to
emerge resplendent on to the savannah immediately after
dropping the 2m from mother to earth. This is of course
the height of laziness compared with something like the
blue whale, which takes less than a year to build a
behemoth – literally; blue whales are 3,000kg at birth, and
almost 8m long. Whether or not the baby is born eyes wide
open and ready to run, or whether it is still helpless as a
literal and proverbial kitten makes a difference – some
mothers, like our old friend Pizza Rat, budget their energy
and invest once the babies are born, carrying heavy slices
of pizza back to the nest so her babies can feast on the
finest-quality milk.

The driver behind the pattern of a species' growth,
however, is balance. The swinging arms of evolutionary
measure have to weigh the need to achieve a certain level
of competency – whether it be hunting lizards or something
else apparently age-linked, like drinking alcohol or driving
a tractor* – against the risk of running out of the energy
needed to fuel that growth. You have to weigh how much
you can feed a baby against how often you would like to
have a new one. Our long human childhoods are literally
fed by the extended care we receive after infancy, in the
period of childhood when we could no more catch a lizard

---

* Or all of these at once if you're on a private road.

than we could distinguish a porter from an IPA or perform any of the other important society-mandated skills.

As we saw throughout this book, after taking longer to get away from mama and her milk, the other great apes hit almost all the major milestones of foraging competence and sexual reproduction at far earlier than we do. We are weaned and off the milk by two or three, hit the first-molar feeding milestone* around seven or eight, and – while puberty and the *capacity* to reproduce comes in around twelve to fifteen years, we don't usually get around to having babies until somewhere around eighteen to *thirty* years. And while we live ever so slightly longer than the chimps, who, to be fair, are ever so slightly smaller, they don't draw a line in the sand halfway through their lives that says 'no children allowed past this point'. The existence of menopause in human females is yet another deviation away from the standard life-history course that has to be explained and the space we've chosen for deviation is the time we take growing up. The high-water marks of our life histories are what separate us from our primate past.

Our species has taken strategies from other mammals – and other primates – and applied them to our own special case. This is most evident in the strange elastic stretching of our life histories, where the traditional markers of birth, reproduction and death have been slid to the extremes of our personal timelines. We are born early, useless and dependent – altricial. We then drag our feet far beyond the normal mammal range of waiting times to

---

* In humans, as in other great apes, this implies the ability to feed oneself. This has very little to do with the lived actuality of the phase of modern childhood where vacuum-packed sachets of cheese tortellini comprises every meal. Which itself can extend far beyond childhood and into the fifth decade.

reproduce, only to then, in a monumental challenge to what almost all other life on earth considers sensible, call time on female reproduction well before we have to. We then stretch these non-reproductive years out, outliving all of our wild primate relatives. There are reasons behind these changes, of course, and they are the things that make us human. Our long lifespans and, particularly, the grandmothers who live them are there to help, not themselves, but us, their descendants. That idiot baby is actually made up of fat and brain, ready to burn an ungodly number of calories because of all the things evolution has forced us to prize – hairlessness, long walks, piña coladas – a baby who learns is the line we actually choose to put our bet on.

We pushed the limits of primate possibility and, what's more, we are still pushing. When it comes to the length of childhood, we are a long-lived species that has slowly and steadily started to behave as though we were an even *longer*-lived species. Take one of the longest-lived animals on the planet, the bowhead whale. These nearly 2-tonne behemoths can live for more than 200 years – as we found out from the rather unexpected discovery of an 1880's explosive harpoon head stuck in a whale that died in 2007. But even these long-lived animals get around to the business of reproduction after a decade and a half, a period where we are still using all of the social means at our disposal to delay the transition to maturity.* We can point to anti-teen-pregnancy campaigns and the precarious structures of our economic worlds as social factors dragging out our long periods of dependence in the modern day, but we still don't know whether our idiosyncratic life histories

---

* No matter how grown up we think we are as 15-year-olds.

are a *fait accompli* set in evolutionary stone – or if we are still adapting them, this far along.

Our species has been having a very, very long conversation with itself these past several hundred thousand years about what a child really is. How old is old enough? And for what? We have seen throughout this book the different approaches to maturation that have carried our species forward, from infants doing not much more than growing brains and making their parents love them to toddlers imposing themselves on social groups and potential allies, older children stopping to stare (and learn) while they work out the business of mastering their physical selves, and then an uncertain amount of time in the active doing of learning to be an adult. This not-quite-grown-up status hangs on until some professional or biological status trumps it: the apprentice becomes the master, the daughter becomes a mother, *etc.* And in the last few chapters we have seen more and more of how our changing lifestyles over the past several thousand years have affected how we spend that long bright teatime of the soul.

The final chapter in the story of our childhood is that of a complicated adolescence that expands or contracts to accumulate the capital that someone – our mother, our father, our family, our friends, our society – is willing to put down on our behalf. We've seen the three forms this can take. A physical investment in the welfare of a child is made arguably from even before conception. This investment is embodied in the physical person of the child – in how tall they grow, how strong or swift they become, how resistant to disease, how free of malnutrition and vitamin deficiencies and, in some cases, whether they live or die. Social capital is the second investment, and covers so much of what we think of as the business of growing up: making friends to play with and learn from; embedding social relations and the skills to maintain them, and giving us a band of supporters if things turn bad. Thirdly, we have the uniquely human potential for investment of actual capital – cold hard cash. We could have an entire additional book-length discussion about whether actual capital is really just a function of physical and social capital that gets ossified in material wealth, but for our purposes the critical point is that when our societies get sufficiently unequal, it takes the ability to leverage things for other things to get your child out of the gutter and into position to survive and thrive. It takes material capital to stretch out a childhood.

In humans, our period of dependency while we master the skills of being human has lengthened and lengthened until it is no longer unusual for a fully mature, breeding-age human to be found pursuing an extended university career rather than, say, the perpetuation of the species. But the world we live in today is profoundly unequal and so are the childhoods within it. Life chances for a child born in poverty in the developing world are astronomically different from that of a child born under brighter, more expensive stars. And life chances do not simply mean the opportunity to rule the world, win a Nobel prize, or make it big as a social media influencer. They include actual chances at life. Simple economic indicators map out killer statistics; if poverty was ever classed as a cause of death it would be number one in the world – in every country. Poor children are more likely to suffer preventable childhood diseases, more likely to have growth falter and to die before they reach adulthood. When it comes to childhood, there is just the one real difference to reckon. It is the one that was there between the snarky schoolboys of Aristophanes and the slaves that served them, and it is still present today, and that is the time we give it. The wealthy, urban and advantaged of our species have long been able to prolong their childhoods for *decades* longer than the least well-off.

We do not have to live this way. As a matter of fact, many societies have put their feet down and demanded that we do better; that we do not let just *some* children be children. Increasingly, we find that society deems it acceptable we are able to offer that advantage to a wider and wider selection of children. This is the promise of early-years care funded by the government, of school meals (and school milk), of free school and apprenticeships for all. While the same societies that give children free milk shudder at the thought of giving adults in need a single penny, most people seem to agree that children should have equal chances in life.* For example, in the UK, as a society, we (well, some of us) agree that it is important to have children who are literate, or numerate, so we collectively invest capital to get every child to that point. We pay taxes and the taxes pay for schools.

And, as a society, we apparently also agree that there is no *benefit* to educating a child beyond what they need to survive. Not to do well, mind you; just to acquire basic skills and even then, perhaps not all of them. As I write, the UK government is ensuring fewer and fewer students attend university to study the arts and humanities, because their complicated finance model loses money on turning out people with a penchant for creative thought. Their point is, how many of us *need* Aristophanes?† We balance our books, we look at the capital returns on our investments, and we judge *these* investments in children to be worthwhile and *these* to be frivolous. When I was a child in California,

---

* But only at the start and only if they're not given too many chances, and also, no free milk.
† Considering that an ability to mangle ancient history is all the qualification you need to run the country, that's a very important question.

society couldn't even come to a consensus on whether or not they were willing to invest in children in general.[*] In 1987, in one of the richest places on the face of the planet, some forty-odd classmates and I shared 15 textbooks that confidently stated Richard Nixon was president.[†]

We look at education, and particularly the long weird adolescence that goes with university education, like it's a privilege that only the finest minds among us deserve – and one that must be paid for in cold, hard cash. There used to be outrage at the idea of educating the poor 'beyond what they would need', and indeed, it's that same sentiment you find undercutting discussions of how we pay for schools today. It is precisely what was said about mandatory, state-funded primary education in the nineteenth century and then again about secondary schooling in the twentieth; one wonders how long they will say it about tertiary, or higher, education at universities. It is certainly said about different groups in society. While education was grudgingly given to poor boys, girls were left out or segregated from many of the institutions societies around the world use to fill up the cultural space between dependence and independence. Children with different social and economic statuses, religions, languages, ethnic backgrounds and genders have all, at various times, been excluded from this great long childhood. At some point, we need to consider whether what we think of as education is really training, or if it is actually just doing the job that the scribal school of Babylon or the one in Aristophanes' Greece was doing: letting the advantaged buy their way into a longer childhood where they benefit from investment and not the other way around. This is a conversation we are still having as a society. My grandmother Doris, who was clever as all get

---

[*] Thatcher took British kids' milk and Reagan took my textbook.
[†] And you wonder why the US has problems?

out, was out of formal education and into a job at the age of 16 and considered a success. Her sons got to university and at least one of their kids got a PhD. It took a mere two generations to give me nearly 25 years of extra adolescence; time to spend getting degrees, getting jobs, writing books, losing jobs and generally elevating my own well-being. Meanwhile, a little girl in Afghanistan will be forced to leave school and childhood behind at the grand old age of eleven.

And that is the thing. Here we are, delaying having babies, living in the interstices between child and adult, for longer and longer every generation. We match this at the other end with a long senescence that is not nearly as senescent as it used to be. Our childhoods are getting longer, even as we push up our life expectancy. You may, particularly if you own a child yourself, look at the 'kids' in their 30s cluttering up their parents' basements, rec rooms and attics, and see a catastrophe, a gutting indictment of the society and economy that doesn't let people grow up as fast as they used to. Or, you could squint a little, and see in that long, long childhood a tiny part of the long, long process of evolution. We may have evolved a brand-new slow form of childhood, but we are humans and we have a choice. We are social animals that make social decisions and the biggest decision we can make is what we *do* with our evolving life histories. Currently, in quite a lot of our societies, we have constructed childhoods that take different shapes; success depends on whether a baby can buy itself more time – or not. The lesson in this book is that our childhoods have become longer, from our hominid ancestors on up, and this is not a bad thing – it allows us time to learn how to be a better monkey. But we, in all our various societies, need to have a long hard think about whether we are really giving everyone the same shot at growing up human.

# Acknowledgements

This book was written in a weird way, at a weird time. I had a baby in a global pandemic, and I wrote a book about never growing up, not necessarily in that order. The world ended, then kept ending, and libraries and my waistline became mere memories, but somehow I still had a manuscript due. I'd like to thank every single person who tried their best to help me through that, from my editors Angelique and of course Jim, all the way through to the nurses and doctors of the UK's National Health Service that kept me from dying (super appreciated). Thanks to all the people who kept me sane by reminding me how to be human: to the virtual voices of the Quiz Crew – Steve, Jess, Kat, Mal, Claire, Gids, Dan & Hana; the long-suffering British Laughers Tarek, Jen, Carrie, Flip, Erin, Sam & Sophie; to Veysel (and to Enkidu and Gilgamesh, who wouldn't have survived without him). Thank you to the Bumps, who reminded me how many ways there are to be a baby: Adelle, Aimee, Brittany, Charlotte, the Katies, Lizzie, Rachel, & Tessa; to my science writer friends at Neuwrite who suffered the early iterations of this book, and my science-fiction writer friends Cat, Elizabeth, Jane, Jess, and Kat who got the flip-side; to my lovely agent Ella; to my TrowelBlazers Becky, Suzie and Tori; and to everyone who ever offered me wine or tsipouro, clandestine or otherwise, while I was trying to finish this damn thing: Rachel and Patrick, Despoina and Polykarpos, Hermione and Alexandros, David and Ewa.

Thank you to the women whose work I hope to God I represented fairly in this book – and it is largely women – may our field some day get over that. Thank you for the most educational coffee chat in the world to the NHM Human Origins team over the years (Chris, Laura, Ali, Silvia, you've got a lot to answer for). Thank you to my Tooth Fairy Crew, Chris D and Louise H, for filling my head with ideas, and especially to Louise for always knowing what's *interesting* about anthropology. Thank you to Andy, for ~~typing the manuscript~~ riding it all out, and to Maeve and now, just in time for the paperback edition, Elliot, who helped the most, with everything.

# Bibliography

## Chapter 1

Borgerhoff Mulder, M., and B. A. Beheim. 'Understanding the Nature of Wealth and Its Effects on Human Fitness.' *Philos Trans R Soc Lond B Biol Sci* 366.1563 (2011): 344–56

Clayton, N. S., and N. J. Emery. 'The Social Life of Corvids.' *Curr Biol* 17 (2007): R652–R56

Kaplan, H. S., and J. Bock. 'Fertility Theory: Embodied-Capital Theory of Life History Evolution.' *International Encyclopedia of the Social & Behavioral Sciences*. Eds Smelser, Neil J. and Paul B. Baltes. Oxford: Pergamon, 2001. 5561–68

Shennan, S. 'Property and Wealth Inequality as Cultural Niche Construction.' *Philos Trans R Soc Lond B Biol Sci* 366.1566 (2011): 918–26

## Chapter 2

Calder, W. A. *Size, Function, and Life History*. Cambridge, MA: Harvard University Press, 1984

Gould, S. J. 'Cope's Rule as Psychological Artefact.' *Nature* 385.6613 (1997): 199–200

Harvey, P. H. 'Life-History Variation: Size and Mortality Patterns.' *Primate Life History and Evolution*. Ed. Rousseau, C. J. New York: Wiley-Liss, 1990. 81–88

Martin, R. D, and A. M. MacLarnon. 'Reproductive Patterns in Primates and Other Mammals: The Dichotomy between Altricial and Precocial Offspring.' *Primate Life History and Evolution*. Ed. DeRousseau, C. J. *Monographs in Primatology* Volume 14. New York: Wiley-Liss, 1990

Sastrawan, W. J. 'The Word 'Orangutan': Old Malay Origin or European Concoction?' *Journal of the Humanities and Social Sciences of Southeast Asia* 176.4 (2020): 532–41

Stearns, S. 'Life-History Tactics: A Review of the Ideas.' *Quart Rev Biol* 51.1 (1976): 3–47

Vadell, M. V., I. E. Gómez Villafañe, and R. Cavia. 'Are Life-History Strategies of Norway Rats (Rattus Norvegicus) and House Mice (Mus Musculus) Dependent on Environmental Characteristics?' *Wildlife Research* 41 (2014): 172–84

## Chapter 3

Arlet, M. E., *et al.* 'Species, Age and Sex Differences in Type and Frequencies of Injuries and Impairments among Four Arboreal Primate Species in Kibale National Park, Uganda.' *Primates* 50.1 (2008): 65

Barrett, J., D. H. Abbott, and L. M. George. 'Extension of Reproductive Suppression by Pheromonal Cues in Subordinate Female Marmoset Monkeys, Callithrix Jacchus.' *Reproduction* 90.2 (1990): 411–18

Bossen, L. 'Toward a Theory of Marriage: The Economic Anthropology of Marriage Transactions.' *Ethnology* 27.2 (1988): 127–44

Clay, Z., *et al.* 'Female Bonobos Use Copulation Calls as Social Signals.' *Biol Lett* 7.4 (2011): 513–16

Dixson, A. F. 'Copulatory and Postcopulatory Sexual Selection in Primates.' *Folia Primatologica* 89.3–4 (2018): 258–86

—. *Primate Sexuality: Comparative Studies of the Prosimians, Monkeys, Apes, and Humans.* Oxford: Oxford University Press, 2012

Domb, L. G., and M. Pagel. 'Sexual Swellings Advertise Female Quality in Wild Baboons.' *Nature* 410.6825 (2001): 204–06

Drea, C. M. 'Bateman Revisited: The Reproductive Tactics of Female Primates.' *Integrative and Comparative Biology* 45.5 (2005): 915–23

Dunbar, R. I. M. *Primates and Their Societies.* Ithaca, NY: Cornell University Press, 1988

Emery Thompson, M. and A. V. Georgiev. 'The High Price of Success: Costs of Mating Effort in Male Primates.' *Int J Primat* 35.3 (2014): 609–27

Fietz, J., *et al.* 'High Rates of Extra-Pair Young in the Pair-Living Fat-Tailed Dwarf Lemur, Cheirogaleus Medius.' *Behav Ecol Sociobiol* 49 (2000): 8–17

Gangestad, S. W., and R. Thornhill. 'Human Oestrus.' *Proc Biol Sci* 275.1638 (2008): 991–1000

Garn, S. M., A. B. Lewis, and R. S. Kerewesky. 'Sex Difference in Tooth Size.' *J Dent Res* 43 (1964): 306

Harcourt, A. H., et al. 'Testis Weight, Body Weight and Breeding System in Primates.' *Nature* 293.5827 (1981): 55–57

Hilgartner, R., et al. 'Determinants of Pair-Living in Red-Tailed Sportive Lemurs (Lepilemur Ruficaudatus).' *Ethology* 118 (2012): 466–79

Kappeler, P. M., and C. P. van Schaik. 'Evolution of Primate Social Systems.' *Int J Primat* 23.4 (2002): 707–40

Key, C., and C. Ross. 'Sex Differences in Energy Expenditure in Non-Human Primates.' *Proc R Soc London Biol Sci* 266.1437 (1999): 2479–85

Lüpold, S., *et al.* 'Sexual Ornaments but Not Weapons Trade Off against Testes Size in Primates.' *Proc R Soc London Biol Sci* 286.1900 (2019): 2018–2542

Matthews, R. 'Storks Deliver Babies (P=0.008).' *Teaching Statistics* 22.2 (2000): 36–38

Parish, A. R. 'Female Relationships in Bonobos (Pan Paniscus).' *Human Nature* (1996): 791–96

Pawłowski, B. 'Loss of Oestrus and Concealed Ovulation in Human Evolution: The Case against the Sexual-Selection Hypothesis.' *Curr Anth* 40.3 (1999): 257–76

Plavcan, J. M. 'Sexual Dimorphism in Primate Evolution.' *Yearbook of Physical Anthropology* 44 (2001): 25–53

Pond, C. M. 'The Significance of Lactation in the Evolution of Mammals.' *Evolution* 31.1 (1977): 177–99

Rigaill, L. 'Multimodal Signalling of Ovulation in Human and Non-Human Primates.' *BMSAP* 26.3 (2014): 161–65

Rode-Margono, E. J., et al. 'The Largest Relative Testis Size among Primates and Aseasonal Reproduction in a Nocturnal Lemur, Mirza Zaza.' *Am J Biol Anth* 158.1 (2015): 165–69

Rooker, K., and S. Gavrilets. 'On the Evolution of Visual Female Sexual Signalling.' *Proc R Soc London Biol Sci* 285.1879 (2018): 20172875

Small, M. F., et al. 'Female Primate Sexual Behavior and Conception: Are There Really Sperm to Spare? [and Comments and Reply].' *Curr Anth* 29.1 (1988): 81–100

Stephen, I. D., et al. 'Skin Blood Perfusion and Oxygenation Colour Affect Perceived Human Health.' *PLoS One* 4.4 (2009): e5083–e83

Swedell, L. 'Primate Sociality and Social Systems.' *Nature Education Knowledge* 3.10 (2012): 84

Szalay, F. S., and R. K. Costello. 'Evolution of Permanent Estrus Displays in Hominids.' *J Hum Evol* 20.6 (1991): 439–64

Trivers, R. L. 'Trivers on Epstein.' Blog (2020)

## Chapter 4

Dunsworth, H., and A. Buchannon. 'I Know Where Babies Come from Therefore I Am Human.' *Aeon* (2019). Web

Fernandez-Duque, E., et al. 'The Evolution of Pair-Living, Sexual Monogamy, and Cooperative Infant Care: Insights from Research on Wild Owl Monkeys, Titis, Sakis, and Tamarins.' *Am J Biol Anth* 171.S70 (2020): 118–73

Fuentes, A. 'Patterns and Trends in Primate Pair Bonds.' *Int J Primat* 23.5 (2002): 953–78

Hrdy, S. B. *The Langurs of Abu: Female and Male Strategies of Reproduction.* Cambridge, MA: Harvard University Press, 1980

Hurlbert, A. C., and Y. Ling. 'Biological Components of Sex Differences in Color Preference.' *Curr Biol* 17.16 (2007): R623–R25

Scelza, B. A. 'Female Choice and Extra-Pair Paternity in a Traditional Human Population.' *Biology Letters* 7.6 (2011): 889–91

Schacht, R., and K. L. Kramer. 'Are We Monogamous? A Review of the Evolution of Pair-Bonding in Humans and Its Contemporary Variation Cross-Culturally.' *Front Ecol Evol* 7.230 (2019)

UNData. 'Live Births by month of birth'. Demographic Statistics Database. United Nations Statistical Division.

## Chapter 5

Altmann, J., et al. 'Body Size and Fatness of Free-Living Baboons Reflect Food Availability and Activity Levels.' *Am J Primat* 30.2 (1993): 149–61

Beehner, J. C., et al. 'The Endocrinology of Pregnancy and Fetal Loss in Wild Baboons.' *Horm Behav* 49.5 (2006): 688–99

Boinski, S. 'Birth Synchrony in Squirrel Monkeys (Saimiri Oerstedi): A Strategy to Reduce Neonatal Predation.' *Behav Ecol Sociobiol* 21.6 (1987): 393–400

Bygdell, M., et al. 'Revisiting the Critical Weight Hypothesis for Regulation of Pubertal Timing in Boys.' *Am J Clin Nutr* 113.1 (2021): 123–28

Chard, T. 'Frequency of Implantation and Early Pregnancy Loss in Natural Cycles.' *Baillieres Clin Obstet Gynaecol* 5.1 (1991): 179–89

Chavez-MacGregor, M., et al. 'Lifetime Cumulative Number of Menstrual Cycles and Serum Sex Hormone Levels in Postmenopausal Women.' *Breast Cancer Res Treat* 108.1 (2008): 101–12

Clancy, K. 'Menstruation Is Just Blood and Tissue You Ended up Not Using.' *Scientific American Blogs* (2011)

Condon, R. G., and R. Scaglion. 'The Ecology of Human Birth Seasonality.' *Hum Ecol* 10.4 (1982): 495–511

Dahlberg, J., and G. Andersson. 'Fecundity and Human Birth Seasonality in Sweden: A Register-Based Study.' *Rep Health* 16.1 (2019): 87

Deschner, T., et al. 'Timing and Probability of Ovulation in Relation to Sex Skin Swelling in Wild West African Chimpanzees, Pan Troglodytes Verus.' *Anim Behav* 66.3 (2003): 551–60

Dittus, W. P. J. 'Arboreal Adaptations of Body Fat in Wild Toque Macaques (Macaca Sinica) and the Evolution of Adiposity in Primates.' *Am J Biol Anth* 152.3 (2013): 333–44

Drea, C. M. 'Bateman Revisited: The Reproductive Tactics of Female Primates.' *Integrat Compar Biol* 45.5 (2005): 915–23

Duan, R., et al. 'The Overall Diet Quality in Childhood Is Prospectively Associated with the Timing of Puberty.' *Eur J Nutr* (2020)

Elks, C. E., *et al.* 'Thirty New Loci for Age at Menarche Identified by a Meta-Analysis of Genome-Wide Association Studies.' *Nature Genetics* 42.12 (2010): 1077–85

Emera, D., *et al.* 'The Evolution of Menstruation: A New Model for Genetic Assimilation.' *BioEssays* 34.1 (2012): 26–35

Frisch, R. E. 'Body Fat, Menarche, Fitness and Fertility.' *Human Reproduction* 2.6 (1987): 521–33

Healy, K., *et al.* 'Animal Life History Is Shaped by the Pace of Life and the Distribution of Age-Specific Mortality and Reproduction.' *Nature Ecology & Evolution* 3.8 (2019): 1217–24

Heldstab, S., *et al.* 'Getting Fat or Getting Help? How Female Mammals Cope with Energetic Constraints on Reproduction.' *Front Zool* 14 (2017): 29–29

Horn, J., and T. J. Vatten. 'Reproductive and Hormonal Risk Factors of Breast Cancer: A Historical Perspective.' *Int J Womens Health* 9 (2017): 265–72

Jauniaux, E., *et al.* 'Placental-Related Diseases of Pregnancy: Involvement of Oxidative Stress and Implications in Human Evolution.' *Hum Reprod Update* 12.6 (2006): 747–55

Kaplowitz, P. B. 'Link between Body Fat and the Timing of Puberty.' *Pediatrics* 121:3 (2008): S208–S17

Knott, C. 'Female Reproductive Ecology of the Apes: Implications for Human Evolution.' Ed. P. T. Ellison. *Reproductive Ecology and Human Evolution* (2011). Hawthorne, New York:

Kühnert, B., and E. Nieschlag. 'Reproductive Functions of the Ageing Male.' *Hum Reprod Update* 10.4 (2004): 327–39

Macklon, N. S., and J. J. Brosens. 'The Human Endometrium as a Sensor of Embryo Quality.' *Biol Reprod* 91.4 (2014): 98

McNeilly, A. S. 'Reproduction and Environment.' *Reproduction and Adaptation: Topics in Human Reproductive Ecology.* Eds Mascie-Taylor, C., G. Nicholas and L. Rosetta. Cambridge Studies in Biological and Evolutionary Anthropology. Cambridge: Cambridge University Press, 2011. 1–16

Miller, G., *et al.* 'Ovulatory Cycle Effects on Tip Earnings by Lap Dancers: Economic Evidence for Human Estrus?' *Evol Hum Behav* 28.6 (2007): 375–81

Nguyen, N. T. K., *et al.* 'Nutrient Intake through Childhood and Early Menarche Onset in Girls: Systematic Review and Meta-Analysis.' *Nutrients* 12.9 (2020)

Pettit, Mi., and J. Vigor. 'Pheromones, Feminism and the Many Lives of Menstrual Synchrony.' *BioSocieties* 10.3 (2015): 271–94

Schwartz, S. M., *et al.* 'Dietary Influences on Growth and Sexual Maturation in Premenarchial Rhesus Monkeys.' *Horm Behav* 22.2 (1988): 231–51

Small, M. F., *et al.* 'Female Primate Sexual Behavior and Conception: Are There Really Sperm to Spare? [and Comments and Reply].' *Curr Anth* 29.1 (1988): 81–100

Strassmann, B. I. 'The Biology of Menstruation in *Homo sapiens*: Total Lifetime Menses, Fecundity, and Nonsynchrony in a Natural-Fertility Population.' *Curr Anth* 38.1 (1997): 123–29

Teklenburg, G., *et al.* 'Natural Selection of Human Embryos: Decidualizing Endometrial Stromal Cells Serve as Sensors of Embryo Quality Upon Implantation.' *PLoS One* 5.4 (2010): e10258

Wang, H., *et al.* 'Maternal Age at Menarche and Offspring Body Mass Index in Childhood.' *BMC Pediatrics* 19.1 (2019): 312

Wilcox, A. J., *et al.* 'Likelihood of Conception with a Single Act of Intercourse: Providing Benchmark Rates for Assessment of Post-Coital Contraceptives.' *Contraception* 63.4 (2001): 211–5

Zihlman, A. L., D. R. Bolter, and C. Boesch. 'Skeletal and Dental Growth and Development In Chimpanzees of the Taï National Park, Côte D'ivoire.' *J Zool* 273 (2007): 63–73

## Chapter 6

Blanco, M. B. 'Reproductive Biology of Mouse and Dwarf Lemurs of Eastern Madagascar, with an Emphasis on Brown Mouse Lemurs (Microcebus Rufus) at Ranomafana National Park, a Southeastern Rainforest.' University of Massachusetts, 2010

Borries, C., *et al.* 'Beware of Primate Life History Data: A Plea for Data Standards and a Repository.' *PLoS One* 8.6 (2013): e67200

Callister, L. C. and I. Khalaf. 'Culturally Diverse Women Giving Birth: Their Stories.' *Childbirth across Cultures: Ideas and Practices of Pregnancy, Childbirth and the Postpartum.* Ed. Selin, Helaine. Dordrecht: Springer Netherlands, 2009. 33–39

Cameron, E. Z., and F. Dalerum. 'A Trivers-Willard Effect in Contemporary Humans: Male-Biased Sex Ratios among Billionaires.' *PLoS One* 4.1 (2009): e4195

Cockburn, A., *et al.* 'Sex Ratios in Birds and Mammals: Can the Hypotheses Be Disentangled?' *Sex Ratios: Concepts and Research Methods.* Ed. Hardy, I. C. W. Cambridge: Cambridge University Press, 2002. 266–86

Dunn-Fletcher, C. E., *et al.* 'Anthropoid Primate–Specific Retroviral Element The1b Controls Expression of Crh in Placenta and Alters Gestation Length.' *PLOS Biology* 16.9 (2018): e2006337

Dunsworth, H., *et al.* 'Metabolic Hypothesis for Human Altriciality.' *PNAS* 109.38 (2012): 15212–16

Fleagle, J. G. 'Chapter 4 - the Prosimians: Lemurs, Lorises, Galagos and Tarsiers.' *Primate Adaptation and Evolution* (Third Edition). Ed. Fleagle, J. G. San Diego: Academic Press, 2013. 57–88

Gatti, S., *et al.* 'Population and Group Structure of Western Lowland Gorillas (Gorilla Gorilla Gorilla) at Lokoué, Republic of Congo.' *Am J Primatol* 63.3 (2004): 111–23

Haig, D. 'Genetic Conflicts in Human Pregnancy.' *Quarter Rev Biol* 68.4 (1993): 495–532

Harvey, P. H., and T. H. Clutton-Brock. 'Life History Variation in Primates.' *Evolution* 39.3 (1985): 559–81

Institute of Medicine (US) Committee on Nutritional Status During Pregnancy and Lactation. *Energy Requirements, Energy Intake, and Associated Weight Gain During Pregnancy*. Washington DC, 1990

Iradukunda, F. 'Food Taboos During Pregnancy.' *Health Care for Women International* 41.2 (2020): 159–68

Lazarus, J. 'Human Sex Ratios: Adaptations and Mechanisms, Problems and Prospects.' *Sex Ratios: Concepts and Research Methods*. Ed. Hardy, I. C. W. Cambridge: Cambridge University Press, 2002. 287–312

Mi, S., *et al.* 'Syncytin Is a Captive Retroviral Envelope Protein Involved in Human Placental Morphogenesis.' *Nature* 403.6771 (2000): 785–89

Ming, C. 'Zhuan Nü Wei Nan Turning Female to Male: An Indian Influence on Chinese Gynaecology?' *Asian Medicine* 1.2 (2005): 315–34

National Research Council (US) Committee on Technological Options to Improve the Nutritional Attributes of Animal Products. *Designing Foods: Animal Product Options in the Marketplace*. Washington DC: National Research Council (US) Committee on Technological Options to Improve the Nutritional Attributes of Animal Products, 1988

Newton-Fisher, N. E. 'The Hunting Behavior and Carnivory of Wild Chimpanzees.' *Handbook of Paleoanthropology*. Eds Henke, W. and I. Tattersall. Berlin, Heidelberg: Springer Berlin Heidelberg, 2015. 1661–91

Peacock, N. R. 'Comparative and Cross-Cultural Approaches to the Study of Human Female Reproductive Failure.' *Primate Life History and Evolution*. Ed. DeRousseau, C. J. New York: Wiley-Liss, 1990. 195–220

Perret, Martine. 'Influence of Social Factors on Sex Ratio at Birth, Maternal Investment and Young Survival in a Prosimian Primate.' *Behav Ecol Sociobiol* 27.6 (1990): 447–54

Sight and Life Foundation. 'Food Taboos During Pregnancy and Lactation across the World.' https://sightandlife.org/wp-content/uploads/2017/02/Food-Taboos-infographic.pdf

Trivers, R. L., and D. E. Willard. 'Natural Selection of Parental Ability to Vary the Sex Ratio of Offspring.' *Science* 179.4068 (1973): 90–2

## Chapter 7

Aiello, L. C., and P. Wheeler. 'The Expensive-Tissue Hypothesis: The Brain and the Digestive System in Human and Primate Evolution.' *Curr Anth* 36.2 (1995): 199–221

Betti, L., and A. Manica. 'Human Variation in the Shape of the Birth Canal Is Significant and Geographically Structured.' *Proc R Soc London Biol Sci* 285.1889 (2018): 20181807

DeSilva, J., and J. Lesnik. 'Chimpanzee Neonatal Brain Size: Implications for Brain Growth in *Homo erectus*.' *J Hum Evol* 51.2 (2006): 207–12

DeSilva, J. M. 'A Shift toward Birthing Relatively Large Infants Early in Human Evolution.' *PNAS* 108.3 (2011): 1022–27

Dobbin, J., and J. Sands. 'Comparative Aspects of the Brain Growth Spurt.' *Early Hum Devel* 311 (1978): 79–83

Dunsworth, H., and L. Eccleston. 'The Evolution of Difficult Childbirth and Helpless Hominin Infants.' *Ann Rev Anthropol* 44.1 (2015): 55–69

Dunsworth, H., *et al.* 'Metabolic Hypothesis for Human Altriciality.' *PNAS* 109.38 (2012): 15212–16

Evans, K. M., and V. J. Adams. 'Proportion of Litters of Purebred Dogs Born by Caesarean Section.' *J Small Animal Practice* 51.2 (2010): 113–18

Hirata, S., *et al.* 'Mechanism of Birth in Chimpanzees: Humans Are Not Unique among Primates.' *Biology Letters* 7.5 (2011): 686–88

Holland, D., *et al.* 'Structural Growth Trajectories and Rates of Change in the First 3 Months of Infant Brain Development.' *JAMA Neurology* 71.10 (2014): 1266–74

Lawn, J. E., *et al.* 'Stillbirths: Rates, Risk Factors, and Acceleration Towards 2030.' *Lancet* 387.10018 (2016): 587–603

Lewin, D., *et al.* '[Contractile Force, Work of the Uterus, Power Developed and Resistance of the Cervix Uteri During Labor. Measurements in Poissy Units].' *J Gynecol Obstet Biol Reprod* (Paris) 5.3 (1976): 333–42

Navarrete, A., *et al.* 'Energetics and the Evolution of Human Brain Size.' *Nature* 480.7375 (2011): 91–93

Pan, W., *et al.* 'Birth Intervention and Non-Maternal Infant-Handling During Parturition in a Nonhuman Primate.' *Primates* 55.4 (2014): 483–88

Pontzer, H., *et al.* 'Metabolic Acceleration and the Evolution of Human Brain Size and Life History.' *Nature* 533.7603 (2016): 390–92

Rosenberg, K. R. 'The Evolution of Modern Human Childbirth.' *Am J Biol Anth* 35.S15 (1992): 89–124

Rosenberg, K., and W. Trevathan. 'Birth, Obstetrics and Human Evolution.' *BJOG: An Int J Gynecol Obstet* 109.11 (2002): 1199–206

Sampson, C. 'Galago Demidoff: Prince Demidoff's Bushbaby.' 2004

Silk, J., et al. 'Gestation Length in Rhesus Macaques (Macaca Mulatta).' *Int J Primat* 14.1 (1993): 95–104

Stoller, M. K. 'The Obstetric Pelvis and Mechanism of Labor in Nonhuman Primates.' University of Chicago, 1995

Tiesler, V. 'Studying Cranial Vault Modifications in Ancient Mesoamerica.' *J Anthropol Sci* 90 (2012): 33–58

Trevathan, W. 'Primate Pelvic Anatomy and Implications for Birth.' *Philos Trans R Soc B Biol Sci* 370.1663 (2015): 20140065

## Chapter 8

Arriaza, B., et al. 'Maternal Mortality in Pre-Columbian Indians of Arica, Chile.' *Am J Biol Anth* 77.1 (1988): 35–41

Campbell, O. 'Why Male Midwives Concealed the Obstetric Forceps.' JSTOR Daily, 2018.

Couto-Ferreira, M. E. 'She Will Give Birth Easily: Therapeutic Approaches to Childbirth in 1st Millennium Bce Cuneiform Sources.' *Dynamis: Acta Hispanica ad Medicinae Scientiarumque Historiam Illustrandam* 34 (2014): 289–315

Drife, J. 'The Start of Life: A History of Obstetrics.' *Postgraduate Medical Journal* 78.919 (2002): 311–15

Fiddyment, S., et al. 'Girding the Loins? Direct Evidence of the Use of a Medieval English Parchment Birthing Girdle from Biomolecular Analysis.' *R Soc Open Sci* 8.3 (2021): 202055

Fitzpatrick-Matthews, K. 'The Woman and Three Babies, the Sad Story of a Real Person.' *North Hertfordshire Museums*. North Hertfordshire Museums, 2020

Green, Monica H. *Women's Healthcare in the Medieval West: Texts and Contexts*. Aldershot Ashgate: Variorum, 2000

Hotelling, B. A. 'From Psychoprophylactic to Orgasmic Birth.' *J Perinat Educ* 18.4 (2009): 45–48

Laudicina, N. M., et al. 'Reconstructing Birth in *Australopithecus sediba*.' *PLoS One* 14.9 (2019): e0221871

Lee, J. 'Childbirth in Early Imperial China.' *NAN NÜ* 7.2 (2005): 216–86

Lieverse, A. R., et al. 'Death by Twins: A Remarkable Case of Dystocic Childbirth in Early Neolithic Siberia.' *Antiquity* 89.343 (2015): 23–38

Loudon, I. 'Deaths in Childbed from the Eighteenth Century to 1935.' *Med Hist* 30.1 (1986): 1–41

Lurie, S. 'Euphemia Maclean, Agnes Sampson and Pain Relief During Labour in 16th Century Edinburgh.' *Anaesthesia* 59.8 (2004): 834–35

Petersen, E. E., *et al.* 'Racial/Ethnic Disparities in Pregnancy-Related Deaths – United States, 2007–2016.' *MMWR Morb Mortal Wkly Rep* 68 (2019): 62–765

Pfeiffer, S., *et al.* 'Discernment of Mortality Risk Associated with Childbirth in Archaeologically Derived Forager Skeletons.' *Int J Paleo* 7 (2014): 15–24

Rosenberg, K., and W. Trevathan. 'Birth, Obstetrics and Human Evolution.' *BJOG: An Int J Gynecol Obstet* 109.11 (2002): 1199–206

Sarkar, S. 'Pregnancy, Birthing, Breastfeeding and Mothering: Hindu Perspectives from Scriptures and Practices.' *Open Theology* 6.1 (2020): 104–16

Simpson, J. Y. 'Discovery of a New Anæsthetic Agent More Efficient Than Sulphuric Æther.' *Provincial Medical and Surgical Journal* (1844–1852) 11.24 (1847): 656–58

Skippen, M., *et al.* 'The Chain Saw - a Scottish Invention.' *Scot Med J* 49.2 (2004): 72–75

—. 'Obstetric Practice and Cephalopelvic Disproportion in Glasgow between 1840 and 1900.' University of Glasgow, 2009

Töpfer, S. 'The Physical Activity of Parturition in Ancient Egypt: Textual and Epigraphical Sources.' *Dynamis: Acta Hispanica ad Medicinae Scientiarumque Historiam Illustrandam* 34 (2014): 317–35

Wegner, J. 'The Magical Birth Brick.' *Expedition Magazine* 48 (2006)

Zhou, Y., *et al.* 'Bioarchaeological Investigation of an Obstetric Death at Huigou Site (3900–2900 Bc), Henan, China.' *Int J Osteo* 30.2 (2020): 264–74

## Chapter 9

Barry, H., and L. M. Paxson. 'Infancy and Early Childhood: Cross-Cultural Codes 2.' *Ethnology* 10.4 (1971): 466–508

Bartick, M., *et al.* 'Babies in Boxes and the Missing Links on Safe Sleep: Human Evolution and Cultural Revolution.' *Mat Child Nutr* 14.2 (2018): e12544

Boswell, J. *The Kindness of Strangers: The Abandonment of Children in Western Europe from Late Antiquity to the Renaissance.* Chicago: University of Chicago Press, 1988

Bourdieu, P. *Distinction: A Social Critique of the Judgement of Taste* Trans. Nice, R. Cambridge, MA: Harvard University Press, 1984

British Museum. 'Tablet; Object 122691.' Online Catalogue. Ed. British Museum. 122691 vols

Carroll, M. 'Archaeological and Epigraphic Evidence for Infancy in the Roman World' *The Oxford Handbook of the Archaeology of*

*Childhood*. Eds Crawford, S., D. M. Hadley and G. Shepherd. Oxford: Oxford University Press, 2018

Crittenden, A. N., *et al*. 'Infant Co-Sleeping Patterns and Maternal Sleep Quality among Hadza Hunter-Gatherers.' *Sleep Health* 4.6 (2018): 527–34

De Lucia, K. 'A Child's House: Social Memory, Identity, and the Construction of Childhood in Early Postclassic Mexican Households.' *Am Anthropol* 112.4 (2010): 607–24

Durband, A. C. 'Artificial Cranial Deformation in Kow Swamp 1 and 5: A Response to Curnoe (2007).' *HOMO* 59.4 (2008): 261–69

Farber, W. 'Magic at the Cradle. Babylonian and Assyrian Lullabies.' *Anthropos* 85.1/3 (1990): 139–48

Foster, C. *Bible Pictures and What They Teach Us* Philadelphia, PA: Foster Publications, 1897

Fruth, B., *et al*. 'Sleep and Nesting Behavior in Primates: A Review.' *Am J Biol Anth* 166.3 (2018): 499–509

Gilmore, H. F., and S. E. Halcrow. 'Sense or Sensationalism? Approaches to Explaining High Perinatal Mortality in the Past.' *Tracing Childhood: Bioarchaeological Investigations of Early Lives in Antiquity*. Eds Thompson, J. L., M. P. Alfonso-Durruty and J. J. Crandall. Florida Press Online: University Press, 2014

Hrdy, Sarah B. 'Comes the Child before Man: How Cooperative Breeding and Prolonged Postweaning Dependence Shaped Human Potentials.' *Hunter-Gatherer Childhoods: Evolutionary, Developmental, and Cultural Perspectives*. Ed. Hewlett, B. New York: Routledge, 2005. 67–91

Lewis, Mary. 'Sticks and Stones: Exploring the Nature and Significance of Child Trauma in the Past.' *The Routledge Handbook of the Bioarchaeology of Human Conflict*. Eds Knüsel, C. and M. Smith. London, UK: Routledge, 2013

McKenna, J. J., *et al*. 'Mother–Infant Cosleeping, Breastfeeding and Sudden Infant Death Syndrome: What Biological Anthropology Has Discovered About Normal Infant Sleep and Pediatric Sleep Medicine.' *Am J Biol Anth* 134.S45 (2007): 133–61

McKenna, J. J., and L. T. Gettler. 'There Is No Such Thing as Infant Sleep, There Is No Such Thing as Breastfeeding, There Is Only Breastsleeping.' *Acta Paediatrica* 105.1 (2016): 17–21

Metropolitan Museum of Art. 'Amulet with a Lamashtu Demon.' Web

Nunn, C. L., and C. P. van Schaik. 'A Comparative Approach to Reconstructing the Socioecology of Extinct Primates.' *Reconstructing Behavior in the Fossil Record*. Eds Plavcan, J. M., *et al*. New York: Kluwer Academic/Plenum, 2002. 159–216

Nunn, C. L., *et al*. 'Shining Evolutionary Light on Human Sleep and Sleep Disorders.' *Evol Med Pub Health* 2016.1 (2016): 227–43

Nunn, C. L., *et al.* 'Primate Sleep in Phylogenetic Perspective. In Evolution of Sleep: Phylogenetic and Functional Perspectives.' Eds McNamara, P., R. A. Barton and C. L. Nunn. Cambridge: Cambridge University Press, 2010

Okumura, M. 'Differences in Types of Artificial Cranial Deformation Are Related to Differences in Frequencies of Cranial and Oral Health Markers in Pre-Columbian Skulls from Peru.' *Boletim do Museu Paraense Emílio Goeldi. Ciências Humanas* 9 (2014): 15–26

Patterson, C. "Not Worth the Rearing": The Causes of Infant Exposure in Ancient Greece.' *Trans Am Philol Assoc* (1974–) 115 (1985): 103–23

Plutarch. 'Chapter 13.' *On Superstition*

Schwartz, J. H., *et al.* 'Two Tales of One City: Data, Inference and Carthaginian Infant Sacrifice.' *Antiquity* 91.356 (2017): 442–54

Sears, W., and M. Sears. *The Baby Book: Everything You Need to Know About Your Baby from Birth to Age Two.* Boston: Little Brown, 1993

Tiesler, V. 'Studying Cranial Vault Modifications in Ancient Mesoamerica.' *J Anthropol Sci* 90 (2012): 33–58

Tomori, C. 'Breastsleeping in Four Cultures: Comparative Analysis of a Biocultural Body Technique.' *Breastfeeding: New Anthropological Approaches.* Eds Tomori, C., A. E. L. Palmquist and E. A. Quinn. Abingdon: Routledge, 2018

## Chapter 10

Amaral, L. Q. 'Mechanical Analysis of Infant Carrying in Hominoids.' *Naturwissenschaften* 95.4 (2008): 281–92

Ballard, O., and A. L. Morrow. 'Human Milk Composition: Nutrients and Bioactive Factors.' *Pediatr Clin North Am* 60.1 (2013): 49–74

Bard, K. 'Primate Parenting.' *Handbook of Parenting: Volume 2: Biology and Ecology of Parenting.* Ed. Bornstein, M. H. London: Lawrence Erlbaum Associates, 2002. 99–140

Cawthon Lang, K. A. 'Bonobo (Pan Paniscus) Conservation.' *Primate Factsheets* (2010). Web

Hahn-Holbrook, J., *et al.* 'Human Milk as "Chrononutrition": Implications for Child Health and Development.' *Pediatric Res* 85.7 (2019): 936–42

Hinde, K. 'Colustrum through a Cultural Lens.' *SPLASH! milk science update* (2017). Web

Hinde, K., and L. A. Milligan. 'Primate Milk: Proximate Mechanisms and Ultimate Perspectives.' *Evol Anthropol* 20.1 (2011): 9–23

Mennella, J. A., and N. K. Bobowski. 'The Sweetness and Bitterness of Childhood: Insights from Basic Research on Taste Preferences.' *Physiol Behav* 152.Pt B (2015): 502–07

Peckre, L., *et al.* 'Holding-On: Co-Evolution between Infant Carrying and Grasping Behaviour in Strepsirrhines.' *Scientific Reports* 6.1 (2016): 37729

Pond, C. M. 'The Significance of Lactation in the Evolution of Mammals.' *Evolution* 31.1 (1977): 177–99

Ramani, S., *et al.* 'Human Milk Oligosaccharides, Milk Microbiome and Infant Gut Microbiome Modulate Neonatal Rotavirus Infection.' *Nat Comms* 9.1 (2018): 5010

Rogers, R. L., and M. Slatkin. 'Excess of Genomic Defects in a Woolly Mammoth on Wrangel Island.' *PLOS Genetics* 13.3 (2017): e1006601

Ross, C. 'Park or Ride? Evolution of Infant Carrying in Primates.' *Int J Primat* 22.5 (2001): 749–71

Stevens, B. J., *et al.* 'Sucrose for Analgesia in Newborn Infants Undergoing Painful Procedures.' *Cochrane Database of Systematic Reviews* 1 (2010): Cd001069

USDA. 'Infant Nutrition and Feeding Guide.' Washington DC, USA: US Department of Agriculture, 2019

Wickes, I. G. 'A History of Infant Feeding. I. Primitive Peoples; Ancient Works; Renaissance Writers.' *Arch Dis Child* 28.138 (1953): 151–8

Zerjal, T., *et al.* 'The Genetic Legacy of the Mongols.' *Am J Hum Gen* 72.3 (2003): 717–21

## Chapter 11

Baitzel, S. I., and P. S. Goldstein. 'More Than the Sum of Its Parts: Dress and Social Identity in a Provincial Tiwanaku Child Burial.' *J Anthropol Arch* 35 (2014): 51–62

Ballard, O, and A. L. Morrow. 'Human Milk Composition: Nutrients and Bioactive Factors.' *Pediatr Clin North Am* 60.1 (2013): 49–74

CDC. 'Breastfeeding Report Card United States, 2020.' Ed. Control, Centers for Disease. Atlanta, GA: Centers for Disease Control and Prevention, 2020

Couto-Ferreira, M. E. 'Being Mothers or Acting (Like) Mothers?' *Women in Antiquity*. Eds Budin, S. L. and J. M. Turfa. London, UK: Taylor & Francis, 2016

Dunne, J., *et al.* 'Milk of Ruminants in Ceramic Baby Bottles from Prehistoric Child Graves.' *Nature* 574.7777 (2019): 246–48

Feucht, E. 'Motherhood in Pharonic Egypt.' *Women in Antiquity*. Eds Budin, S. L. and J. M. Turfa. London, UK: Taylor & Francis, 2016

Fildes, V. 'The English Wet-Nurse and Her Role in Infant Care 1538–1800.' *Med Hist* 32.2 (1988): 142–73

Killgrove, K. 'Where Did Ancient Roman Babies Poop?' *Forbes* (2017)

Lacaille, A. D. 'Infant Feeding-Bottles in Prehistoric Times.' *Proc R Soc Med* 43.7 (1950): 565–8

Lynch, K. M., and J. K. Papadopoulos. 'Sella Cacatoria: A Study of the Potty in Archaic and Classical Athens.' *Hesperia* 75.1 (2006): 1–32

Mair, V. H. 'Ancient Mummies of the Tarim Basin.' *Expedition Magazine* 58.2 (2016)

Miller, M. *The Baby Killer.* London, UK: War on Want, 1974

Moore, E. R., *et al.* 'Early Skin-to-Skin Contact for Mothers and Their Healthy Newborn Infants.' *Cochrane Database of Systematic Reviews* 11 (2016)

Morgan, J. L. '"Some Could Suckle over Their Shoulder": Male Travelers, Female Bodies, and the Gendering of Racial Ideology, 1500–1770.' *The William and Mary Quarterly* 54.1 (1997): 167–92

Mulkerin, M. 'Seal Skin Baby Pants and Ancient Diapers.' *Rogers Archaeology Lab* (2014)

Rhodes, M. C. 'Domestic Vulnerabilities: Reading Families and Bodies into Eighteenth-Century Anglo-Atlantic Wet Nurse Advertisements.' *Journal of Family History* 40.1 (2014): 39–63

Scelza, B. A., and K. Hinde. 'Crucial Contributions.' *Human Nature* 30.4 (2019): 371–97

Stevens, E. E., *et al.* 'A History of Infant Feeding.' *J Perinat Educ* 18.2 (2009): 32–39

Tessier, R., *et al.* 'Kangaroo Mother Care: A Method for Protecting High-Risk Low-Birth-Weight and Premature Infants against Developmental Delay.' *Infant Behavior and Development* 26.3 (2003): 384–97

Toth, P. 'Children in Ancient Egypt.' British Museum

UNICEF. 'Breastfeeding: A Mother's Gift, for Every Child.' New York: UNICEF Nutritional Division, 2018

Victora, C. G., *et al.* 'Breastfeeding in the 21st Century: Epidemiology, Mechanisms, and Lifelong Effect.' *Lancet* 387.10017 (2016): 475–90

Volk, A. A. 'Human Breastfeeding Is Not Automatic: Why That's So and What It Means for Human Evolution.' *J Social Evol Cultural Psychol* 3.4 (2009): 305–14

Wegner, J. 'A Decorated Birth-Brick from South Abydos: New Evidence on Childbirth and Birth Magic in the Middle Kingdom'. *Archaism and Innovation: Studies in the Culture of Middle Kingdom Egypt.* Eds. D. P. Silverman et al. New Haven, CT / Philadelphia PA: Department of Near Eastern Languages and Civilizations, Yale University / University of Pennsylvania Museum, 2009. 447–496

West, E., and R. J. Knight. 'Mothers' Milk: Slavery, Wet-Nursing, and Black and White Women in the Antebellum South.' *J Southern Hist* 83 (2017): 37–68

WHO. 'Exclusive Breastfeeding under 6 Months.' (2019)

Williamson, I., *et al.* '"It Should Be the Most Natural Thing in the World": Exploring First Time Mothers' Breastfeeding Difficulties in the UK Using Audio-Diaries and Interviews.' *Mat Child Nutr* 8 (2012): 434–47

Xu, F., *et al.* 'Breastfeeding in China: A Review.' *International Breastfeeding Journal* 4.1 (2009): 6

## Chapter 12

Barry, H., and L. M. Paxson. 'Infancy and Early Childhood: Cross-Cultural Codes 2.' *Ethnology* 10.4 (1971): 466–508

Borráz-León, J. I., *et al.* 'Low Intrasexual Competitiveness and Decreasing Testosterone in Human Males (*Homo sapiens*): The Adaptive Meaning.' *Behaviour* 157.1 (2019): 1

Darwin, C. *Descent of Man.* 1871 (1981) DeGrutyer.

Fernandez-Duque, E., *et al.* 'The Evolution of Pair-Living, Sexual Monogamy, and Cooperative Infant Care: Insights from Research on Wild Owl Monkeys, Titis, Sakis, and Tamarins.' *Am J Biol Anth* 171.S70 (2020): 118–73

Fromhage, L. 'Parental Care and Investment.' Els. Wiley Online Library. 1–7

Geniole, S. N., *et al.* 'Is Testosterone Linked to Human Aggression? A Meta-Analytic Examination of the Relationship between Baseline, Dynamic, and Manipulated Testosterone on Human Aggression.' *Horm Behav* (2019): 104644

Gimbutas, M. *Civilization of the Goddess.* San Francisco: Harper, 1991

Harris, R. A., *et al.* 'Evolutionary Genetics and Implications of Small Size and Twinning in Callitrichine Primates.' *PNAS* 111.4 (2014): 1467–72

Hrdy, S. B. 'Cooperative Breeding and the Paradox of Facultative Fathering.' *Family, Ties and Care: Family Transformation in a Plural Modernity.* Eds Bertram, H. and N. Ehlert, 2018. 207

Kaplan, H. S., and J. B. Lancaster. 'An Evolutionary and Ecological Analysis of Human Fertility, Mating Patterns, and Parental Investment.' *Offspring: Human Fertility Behavior in Biodemographic Perspective.* Eds Wachter, K. W. and Bulatao, R. A. Washington DC: National Academies Press, 2003

Kleiman, D. G., and J. R. Malcolm. 'The Evolution of Male Parental Investment in Mammals.' *Parental Care in Mammals.* Eds Gubernick, J. and P. H. Klopfer. Boston, MA: Springer US, 1981. 347–87

Kokko, H. 'Parental Effort and Investment.' *The International Encyclopedia of Anthropology.* Ed. Callan, H.: Wiley Online Library, 2020. 1–7

Maestripieri, D. 'Infant Kidnapping among Group-Living Rhesus Macaques: Why Don't Mothers Rescue Their Infants?' *Primates* 34.2 (1993): 211–16

Malinowski, B. *The Family among the Australian Aborigines: A Sociological Study*. London: University of London Press 1913

Murray, C. M., *et al*. 'Chimpanzee Fathers Bias Their Behaviour Towards Their Offspring.' *R Soc Open Sci* 3.11 (2016): 160441–41

Opie, C., *et al*. 'Male Infanticide Leads to Social Monogamy in Primates.' *PNAS* 110.33 (2013): 13328–32

Sandel, A. A., *et al*. 'Paternal Kin Discrimination by Sons in Male Chimpanzees Transitioning to Adulthood'. *bioArxiv preprint (non peer-reviewed)* https://doi.org/10.1101/631887 (2020)

Sperling, S. 'Baboons with Briefcases: Feminism, Functionalism, and Sociobiology in the Evolution of Primate Gender.' *Signs* 17.1 (1991): 1–27

Storey, A. E., and T. E. Ziegler. 'Primate Paternal Care: Interactions between Biology and Social Experience.' *Horm Behav* 77 (2016): 260–71

Trivers, R. L. 'Parental Investment and Sexual Selection.' Sexual Selection and the Descent of Man, 1871–1971. Ed. Campbell, B. Chicago, IL: Aldine, 1972

Weimerskirch, H., *et al*. 'Sex Differences in Parental Investment and Chick Growth in Wandering Albatrosses: Fitness Consequences.' *Ecology* 81.2 (2000): 309–18

## Chapter 13

Ballard, O., and A. L. Morrow. 'Human Milk Composition: Nutrients and Bioactive Factors.' *Pediatr Clin North Am* 60.1 (2013): 49–74

Cawthon Lang, K. A. 'Primate Factsheets: Gorilla (Gorilla) Behavior.' (2005). Web

CIA. 'Mother's Mean Age at First Birth.' World Factbook

Cloutier, C., *et al*. 'Age-Related Decline in Ovarian Follicle Stocks Differ between Chimpanzees (Pan Troglodytes) and Humans.' *Age (Dordr)* 37.1 (2015): 9746–46

Ellis, S., *et al*. 'Postreproductive Lifespans Are Rare in Mammals.' *Ecology and Evolution* 8.5 (2018): 2482–94

Faddy, M. J., *et al*. 'Accelerated Disappearance of Ovarian Follicles in Mid-Life: Implications for Forecasting Menopause.' *Hum Reprod* 7.10 (1992): 1342–6

Hawkes, K., *et al*. 'Hardworking Hadza Grandmothers.' *Comparative Socioecology*. Eds Standen, V. and R. A. Foley. Oxford: Blackwell Scientific Press, 1989. 341–66

Hawkes, K., *et al*. 'Grandmothering, Menopause, and the Evolution of Human Life Histories.' *PNAS* 95.3 (1998): 1336–39

Herndon, J. G., et al. 'Menopause Occurs Late in Life in the Captive Chimpanzee (Pan Troglodytes).' *Age (Dordr)* 34.5 (2012): 1145–56

Hrdy, S. B. *The Woman That Never Evolved.* Cambridge, MA: Harvard University Press, 1981

McNeilly, A. S. 'Lactation and Fertility.' *Journal of Mammary Gland Biology and Neoplasia* 2.3 (1997): 291–98

Office for National Statistics. 'Marriages in England and Wales: 2017.' (2020)

Stansfield, F. J., et al. 'The Progression of Small-Follicle Reserves in the Ovaries of Wild African Elephants (Loxodonta Africana) from Puberty to Reproductive Senescence.' *Repro Fertil Devel* 25.8 (2013): 1165–73

Wood, B., et al. 'Demographic and Hormonal Evidence for Menopause in Wild Chimpanzees.' *Science* 382.6669 (2023): 368–369

Woods, D. C., et al. 'Oocyte Family Trees: Old Branches or New Stems?' *PLOS Genetics* 8.7 (2012). e1002848

World Health Organisation. 'World Fertility Data 2008.' (2008). Web

World Wildlife Fund. 'Orangutan Factsheet.'

## Chapter 14

Bogin, B., et al. *Human Biology: An Evolutionary and Biocultural Perspective.* Hoboken, NJ: John Wiley & Sons, 2012

Brimacombe, C. S. 'The Enigmatic Relationship between Epiphyseal Fusion and Bone Development in Primates.' *Evol Anthropol* 26.6 (2017): 325–35

Cawthon Lang, K. A. 'Bonobo (Pan Paniscus) Conservation.' *Primate Factsheets* (2010). Web

—. 'Chimpanzee (Pan Troglodytes) Behavior.' *Primate Factsheets* (2006)

Dunsworth, H. 'Expanding the Evolutionary Explanations for Sex Differences in the Human Skeleton.' *Evol Anthropol* 29 (2020) 108–116.

Foerster, S., et al. 'Seasonal Energetic Stress in a Tropical Forest Primate: Proximate Causes and Evolutionary Implications.' *PLoS One* 7.11 (2012): e50108

Heldstab, S. A., et al. 'Reproductive Seasonality in Primates: Patterns, Concepts and Unsolved Questions.' *Biol Reviews* 96.1 (2021): 66–88

Plavcan, J. M. 'Sexual Dimorphism in Primate Evolution.' *Yearbook of Physical Anthropology* 44 (2001): 25–53

Sarringhaus, L. A., et al. 'Locomotor and Postural Development of Wild Chimpanzees.' *J Hum Evol* 66 (2014): 29–38

Schultz, A. H. *The Life of Primates.* New York: Universe Press, 1969

Smith, B. H. 'Life History and the Evolution of Human Maturation.' *Evol Anthropol* 1.4 (1992): 134–42

Swanson, E. M., *et al.* 'Ontogeny of Sexual Size Dimorphism in the Spotted Hyena (Crocuta Crocuta).' *J Mammol* 94.6 (2013): 1298–310

Young, J. W., and L. J. Shapiro. 'Developments in Development: What Have We Learned from Primate Locomotor Ontogeny?' *Am J Biol Anth* 165.S65 (2018): 37–71

Zihlman, A. L., *et al.* 'Skeletal and Dental Growth and Development In Chimpanzees of the Taï National Park, Côte D'ivoire.' *J Zool* 273 (2007): 63–73

## Chapter 15

Bolter, D. R., *et al.* 'Immature Remains and the First Partial Skeleton of a Juvenile *Homo naledi*, a Late Middle Pleistocene Hominin from South Africa.' *PLoS One* 15.4 (2020): e0230440

Brimacombe, C. S. 'The Enigmatic Relationship between Epiphyseal Fusion and Bone Development in Primates.' *Evol Anthropol* 26.6 (2017): 325–35

Cameron, N., *et al.* 'The Postcranial Skeletal Maturation of *Australopithecus sediba*.' *Am J Biol Anth* 163.3 (2017): 633–40

De Groote, I., *et al.* 'New Genetic and Morphological Evidence Suggests a Single Hoaxer Created "Piltdown Man".' *R Soc Open Sci* 3.8 (2016): 160328

Dean, M. C., and B. H. Smith. 'Growth and Development of the Nariokotome Youth, Knm-Wt 15000.' *The First Humans – Origin and Early Evolution of the Genus Homo: Contributions from the Third Stony Brook Human Evolution Symposium and Workshop October 3 – October 7, 2006.* Eds Grine, F. E., J. G. Fleagle and R. E. Leakey. Dordrecht: Springer Netherlands, 2009. 101–20

Garvin, H. M., *et al.* 'Body Size, Brain Size, and Sexual Dimorphism in *Homo naledi* from the Dinaledi Chamber.' *J Hum Evol* 111 (2017): 119–38

Leigh, S. R. 'Evolution of Human Growth.' *Evol Anthropol* 10.6 (2001): 223–36

Lordkipanidze, D., *et al.* 'Postcranial Evidence from Early *Homo* from Dmanisi, Georgia.' *Nature* 449.7160 (2007): 305–10

Prüfer, K., *et al.* 'The Bonobo Genome Compared with the Chimpanzee and Human Genomes.' *Nature* 486.7404 (2012): 527–31

Rizal, Y., *et al.* 'Last Appearance of *Homo erectus* at Ngandong, Java, 117,000–108,000 years Ago.' *Nature* 577.7790 (2020): 381–85

Sutikna, T., *et al.* 'The Spatio-Temporal Distribution of Archaeological and Faunal Finds at Liang Bua (Flores, Indonesia) in Light of the Revised Chronology for *Homo floresiensis*.' *J Hum Evol* 124 (2018): 52–74

Welker, F., *et al.* 'The Dental Proteome of *Homo antecessor*.' *Nature* 580.7802 (2020): 235–38

# Chapter 16

AlQahtani, S. J. *et al.* 'Brief Communication: The London Atlas of Human Tooth Development and Eruption.' *American Journal of Physical Anthropology* 142.3 (2010): 481–90

Aristotle. *Aristotle's History of Animals in Ten Books*. Trans. Cresswell, Richard. London: Geroge Bell and Sons / Project Gutenburg, 1887

Beynon, A. D., and M. C. Dean. 'Distinct Dental Development Patterns in Early Fossil Hominids.' *Nature* 335.6190 (1988): 509–14

Conroy, G. C., and M. W. Vannier. 'The Nature of Taung Dental Maturation Continued.' *Nature* 333.6176 (1988). 808–08

Dean, M. C. 'Growing up Slowly 160,000 Years Ago.' *PNAS* 104.15 (2007): 6093

Dean, M. C., and B. H. Smith. 'Growth and Development of the Nariokotome Youth, Knm-Wt 15000.' *The First Humans – Origin and Early Evolution of the Genus Homo: Contributions from the Third Stony Brook Human Evolution Symposium and Workshop October 3 – October 7, 2006*. Eds Grine, F. E., J. G. Fleagle and R. E. Leakey. Dordrecht: Springer Netherlands, 2009. 101–20

Fatemifar, G., *et al.* 'Genome-Wide Association Study of Primary Tooth Eruption Identifies Pleiotropic Loci Associated with Height and Craniofacial Distances.' *Hum Molec Gen* 22.18 (2013): 3807–17

Humphrey, L., and C. Stringer. *Our Human Story: Where We Come from and How We Evolved*. London: Natural History Museum, 2018

Humphrey, L. 'Weaning Behaviour in Human Evolution.' *Seminars in Cell & Developmental Biology* 21.4 (2010): 453–61

Liversidge, H. 'Variation in Modern Human Dental Development'. *Patterns of Growth and Development in the Genus* Homo. Eds. Nelson, A. J. et al. Cambridge: Cambridge University Press, 2003. 73–113

Mahoney, P. *et al.* 'Growth of Neanderthal Infants from Krapina (120 ~ 130 Ka), Croatia'. *Proc Royal Soc B: Biol Sci* 288.1963 (2021): 20212079

Mann, A. 'The Nature of Taung Dental Maturation.' *Nature* 333.6169 (1988): 123–23

Meyer, M., *et al.* 'A High-Coverage Genome Sequence from an Archaic Denisovan Individual.' *Science* 338.6104 (2012): 222–26

Modesto-Mata, M. *et al.* 'Early and Middle Pleistocene Hominins from Atapuerca (Spain) Show Differences in Dental Developmental Patterns'. *Am J Biol Anth* (2022): 1–13

Ramirez R., *et al.* 'Surprisingly Rapid Growth in Neanderthals.' *Nature* 428.6986 (2004): 936–9

Schour, I., and M. Massler. 'The Development of the Human Dentition.' *J Am Dent Assoc* 28 (1941): 153–60

Smith, B. H. 'Dental Development and the Evolution of Life History in Hominidae.' *Am J Biol Anth* 86.2 (1991): 157–74

Smith, B. H., *et al.* 'Ages of Eruption of Primate Teeth: A Compendium for Aging Individuals and Comparing Life Histories.' *Am J Biol Anth* 37.S19 (1994): 177–231

Smith, T. M., *et al.* 'Earliest Evidence of Modern Human Life History in North African Early *Homo sapiens*.' *PNAS* 104.15 (2007): 6128–33

Tobias, P. V. 'When and by Who Was the Taung Skill Discovered?' *Festschrift for Santiago Genovés*. Ed. Tapa, L. L. Mexico D. F.: Instituto de Investigaciones Antropologicas, Universidad Autonoma de Mexico, 1990. 207–14

Xing, S., *et al.* 'First Systematic Assessment of Dental Growth and Development in an Archaic Hominin (Genus *Homo*) from East Asia.' *Science Advances* 5.1 (2019): eaau0930

## Chapter 17

Guemple, L. 'Teaching Social Relations to Inuit Children.' *Hunters and Gatherers 2: Property, Power and Ideology*. Eds Ingold, T., D. Riches and J. Woodburn. Oxford: Berg Publishers Ltd, 1988. 131–49

Lonsdorf, E. V. 'What Is the Role of Mothers in the Acquisition of Termite-Fishing Behaviors in Wild Chimpanzees (Pan Troglodytes Schweinfurthii)?' *Animal Cog* 9.1 (2006): 36–46

Lonsdorf, E. V., *et al.* 'Sex Differences in Learning in Chimpanzees.' *Nature* 428.6984 (2004): 715–16

MacDonald, K. 'Cross-Cultural Comparison of Learning in Human Hunting.' *Hum Nature* 18.4 (2007): 386–402

Matsuzawa, T. 'Hot-Spring Bathing of Wild Monkeys in Shiga-Heights: Origin and Propagation of a Cultural Behavior.' *Primates* 59.3 (2018): 209–13

——. 'Sweet-Potato Washing Revisited: 50th Anniversary of the Primates Article.' *Primates* 56.4 (2015): 285–87

Onyango, P. O., *et al.* 'Puberty and Dispersal in a Wild Primate Population.' *Horm Behav* 64.2 (2013): 240–49

Ottoni, E. B., *et al.* 'Watching the Best Nutcrackers: What Capuchin Monkeys (Cebus Apella) Know About Others' Tool-Using Skills.' *Anim Cogn* 8.4 (2005): 215–9

Pruetz, J. D., and P. Bertolani. 'Savanna Chimpanzees, Pan Troglodytes Verus, Hunt with Tools.' *Curr Biol* 17.5 (2007): 412–17

Rajpurohit, L. P., and V. Sommer. 'Juvenile Male Emmigration from One-Male Natal Troops in Hanuman Langurs.' *Juvenile Primates: Life History, Development, and Behavior.* Eds Periera, M. and L. A. Fairbanks. Chicago: University of Chicago Press, 2002. 86–103

Watts, D., and A. E. Pusey. 'Behavior of Juvenile and Adolescent Great Apes.' *Juvenile Primates: Life History, Development, and Behavior.* Eds Periera, M. and L. A. Fairbanks. Chicago: University of Chicago Press, 2002. 148–67

Whiten, A., and E. van de Waal. 'The Pervasive Role of Social Learning in Primate Lifetime Development.' *Behav Ecol Sociobiol* 72.5 (2018): 80–80

Ziegler, M., *et al.* 'Development of Middle Stone Age Innovation Linked to Rapid Climate Change.' *Nat Comms* 4.1 (2013): 1905

## Chapter 18

Azéma, M., and F. Rivère. 'Animation in Palaeolithic art: a pre-echo of cinema.' *Antiquity* 86.332 (2012): 316–24

Brumm, A., *et al..* 'Oldest cave art found in Sulawesi.' *Science Advances* 7.3 (2021): 46–8

Conard, N. J. 'A female figurine from the basal Aurignacian of Hohle Fels Cave in southwestern Germany.' *Nature* 459. 7244 (2009): 248–52

Cormier, L. 'Animism, cannibalism, and pet-keeping among the Guajá of Eastern Amazonia.' *Tipiti* 1 (2003): 71–88

De Lucia, K. 'A Child's House: Social Memory, Identity, and the Construction of Childhood in Early Postclassic Mexican Households.' *Am Anthropol* 112.4 (2010): 607–24

Finkel, I. 'Ancient board games in perspective: papers from the 1990 British Museum colloquium with additional contributions.' London: British Museum Press, 2007

Fouts, H. N., *et al.* 'Gender Segregation in Early-Childhood Social Play among the Bofi Foragers and Bofi Farmers in Central Africa.' *Am J Play* 5.3: 333–56.

Garcia, M. A. 'Ichnologie générale de la grotte Chauvet.' *Bulletin de la Société préhistorique française 102 (La grotte Chauvet à Vallon-Pont-d'Arc: un bilan des recherches pluridisciplinaires Actes de la séance de la Société préhistorique française 11 et 12 Octobre 2003, Lyon)* (2005): 103–8

Huffman, M. A., *et al.* 'Cultured Monkeys: Social Learning Cast in Stones.' *Curr Direct Psych Sci.* 17.6 (2008): 410

Langley, M. C. 'Magdalenian Children: Projectile Points, Portable Art and Playthings.' *Ox J Arch* 37.1 (2018): 3–24

Lew-Levy, S., *et al*. 'Gender-Typed and Gender-Segregated Play Among Tanzanian Hadza and Congolese BaYaka Hunter-Gatherer Children and Adolescents.' *Child Dev*, 91 (2019): 1284–1301

Montgomery, S. H. 'The relationship between play, brain growth and behavioural flexibility in primates.' *Anim Behav* 90 (2014): 281–86.

Muratov, M. B. *Greek Terracotta Figurines with Articulated Limbs*. New York: Metropolitan Museum of Art, 2004

Palagi, E. 'Not just for fun! Social play as a springboard for adult social competence in human and non-human primates.' *Behav Ecol Sociobiol* 72.6 (2018): 90.

Varma, S. *Material Culture and Childhood in Harappan South Asia*. Eds Crawford, S., D. M. Hadley and G. Shepherd. Oxford: Oxford University Press, 2018

## Chapter 19

—— 'Civita Giuliana – the Vault of a Cryptoporticus in the Villa Emerges from the New Excavations.' Pompeii, 2020

Avramidou, A. 'Women Dedicators on the Athenian Acropolis and Their Role in Family Festivals: The Evidence for Maternal Votives between 530–450 Bce.' *Cahiers « Mondes Anciens » 6. Les mères et le politique* (2015): 1–29

Benefiel, R. R. 'The Culture of Writing Graffiti within Domestic Spaces at Pompeii.' *Inscriptions in the Private Sphere in the Greco-Roman World*. Eds Benefiel, R. and P. Keegan. Leiden, The Netherlands: Brill, 2016. 80–110

Cooney, K. 'Apprenticeship and Figured Ostraca from the Ancient Egyptian Village of Deir El-Medina.' *Archaeology and Apprenticeship: Body Knowledge, Identity, and Communities of Practice*. Ed. Wendrich, W. Tuscon, AZ: Arizona State University Press, 1982

Crown, P. L. 'Learning to Make Pottery in the Prehispanic American Southwest.' *J Anthropol Res* 57.4 (2001): 451–69

—. 'Life Histories of Pots and Potters: Situating the Individual in Archaeology.' *Am Antiquity* 72.4 (2007): 677–90

Dorland, S. G. H. 'The Touch of a Child: An Analysis of Fingernail Impressions on Late Woodland Pottery to Identify Childhood Material Interactions.' *JAS Reports* 21 (2018): 298–304

Foster, K., *et al*. 'Texts, Storms, and the Thera Eruption.' *J Near Eastern Stud* 55.1 (1996): 1–14

Gelb, I. J. 'The Arua Institution.' *Revue d'Assyriologie et d'archéologie orientale* 66.1 (1972): 1–32

Harrington, N. 'A World without Play?: Children in Ancient Egyptian Art and Iconography.' *The Oxford Handbook of the Archaeology of Childhood*. Eds Crawford, S., D. M. Hadley and G. Shepherd. Oxford: Oxford University Press, 2018

Huntley, K. V. 'Children's Graffiti in Roman Pompeii and Herculaneum.' *The Oxford Handbook of the Archaeology of Childhood*. Eds Crawford, S., D. M. Hadley and G. Shepherd. Oxford: Oxford University Press, 2018

Kramer, K. L. 'Children's Help and the Pace of Reproduction: Cooperative Breeding in Humans.' *Evol Anthropol* 14.6 (2005): 224–37

Kramer, K. L., and J. L. Boone. 'Why Intensive Agriculturalists Have Higher Fertility: A Household Energy Budget Approach.' *Curr Anth* 43.3 (2002): 511–17

LaMoreaux, P. E. 'Worldwide Environmental Impacts from the Eruption of Thera.' *Environ Geol* 26.3 (1995): 172–81

Lee, R. B. 'What Hunters Do for a Living, or, How to Make out on Scarce Resources.' *Man the Hunter*. Eds Lee, R. B. and I. Devore. New York: Routledge, 1968

Manning, S. W. 'Eruption of Thera/Santorini.' *The Oxford Handbook of the Bronze Age Aegean*. Ed. Cline, E. H. Oxford: Oxford University Press, 2012

Marshall, A. *Être Un Enfant En Égypte Ancienne*. Monaco: Rocher, 2013

McCorriston, J. 'The Fiber Revolution: Textile Extensification, Alienation, and Social Stratification in Ancient Mesopotamia.' *Curr Anth* 38.4 (1997): 517–35

Pany-Kucera, D., *et al.* 'Children in the Mines? Tracing Potential Childhood Labour in Salt Mines from the Early Iron Age in Hallstatt, Austria.' *Child Past* 12.2 (2019): 67–80

Rehak, P. 'Children's Work: Girls as Acolytes in Aegean Ritual and Cult.' *Hesperia Supplements* 41 (2007): 205–25

Sahlins, M. 'The Original Affluent Society.' *Stone Age Economics*. New York: Routledge, 1974

## Chapter 20

Al-Rashid, M. '"Schoolboy, Where Have You Been Going So Long?": The Old Babylonian Student and School.' *Everyday Stories from the Ancient Past*, 2019

Aristophanes. 'The Archanians, Knights, Clouds, Wasps, Peace, and Birdsons.' Trans. Hicke, W. J. *The Comedies of Aristophanes, a new and literal translation from the revised text of Dindorf with notes and extracts from the best metrical versions.* London: George Bell & Sons, 1901

Bhutta, Z. A., *et al.* 'Countdown to 2015 Decade Report (2000–10): Taking Stock of Maternal, Newborn, and Child Survival.' *Lancet* 375.9730 (2010): 2032–44

Chasse, G. 'Investigations of Possible Cases of Scurvy in Juveniles from the Kellis 2 Cemetery in the Dakhleh Oasis, Egypt, through Stable Carbon and Nitrogen Isotopic Analysis of Multiple Tissues.' University of Central Florida, 2018

Grosman, L., and N. D. Munro. 'A Natufian Ritual Event.' *Curr Anth* 57.3 (2016): 311–31

Kedar, S. 'La Famille Dans Le Proche-Orient Ancien: Réalités, Symbolismes Et Images.' *Apprenticeship in the Neo-Babylonian Period: A Study of Bargaining Power.* Ed. Lionel, M. Philadelphia, PA: Penn State University Press, 2021. 537–46

Klaus, H. D. 'Subadult Scurvy in Andean South America: Evidence of Vitamin C Deficiency in the Late Pre-Hispanic and Colonial Lambayeque Valley, Peru.' *Int J Paleo* 5 (2014): 34–45

Nissen, H. J. 'The Emergence of Writing in the Ancient near East.' *Interdisciplinary Science Reviews* 10.4 (1985): 349–61

Orschiedt, J. 'The Late Upper Palaeolithic and Earliest Mesolithic Evidence of Burials in Europe.' *Philos Trans R Soc B Biol Sci* 373.1754 (2018): 20170264

Pitre, M. C., *et al.* 'First Probable Case of Scurvy in Ancient Egypt at Nag El-Qarmila, Aswan.' *Int J Paleo* 13 (2016): 11–19

Pritchard, J. *The Ancient near East: An Anthology of Texts and Pictures.* Princeton: Princeton University Press, 1975

Schug, G. R., and K. E. Blevins. 'The Center Cannot Hold.' *A Companion to South Asia in the Past* (2016): 255–73

Turner, B. L., and G. J. Armelagos. 'Diet, Residential Origin, and Pathology at Machu Picchu, Peru.' *Am J Biol Anth* 149.1 (2012): 71–83

Veenhof, K. R. *Letters in the Louvre.* Leiden, The Netherlands: Brill, 2005

Zhang, H., *et al.* 'Osteoarchaeological Studies of Human Systemic Stress of Early Urbanization in Late Shang at Anyang, China.' *PLoS One* 11.4 (2016): e0151854

## Chapter 21

George, J. C., *et al.* 'Age and Growth Estimates of Bowhead Whales (Balaena Mysticetus) Via Aspartic Acid Racemization.' *Canadian J Zool* 77 (2007): 571–80

# Index